SPORTS CAR COLOR HISTORY
MAZDA RX-7

John Matras

MBI Publishing Company

First published in 1994 by MBI Publishing Company, PO Box 1, 729 Prospect Avenue, Osceola, WI 54020-0001 USA

© John Matras, 1994

All rights reserved. With the exception of quoting brief passages for the purposes of review no part of this publication may be reproduced without prior written permission from the Publisher.

The information in this book is true and complete to the best of our knowledge. All recommendations are made without any guarantee on the part of the author or Publisher, who also disclaim any liability incurred in connection with the use of this data or specific details.

We recognize that some words, model names and designations, for example, mentioned herein are the property of the trademark holder. We use them for identification purposes only. This is not an official publication.

MBI Publishing Company books are also available at discounts in bulk quantity for industrial or sales-promotional use. For details write to Special Sales Manager at Motorbooks International Wholesalers & Distributors, 729 Prospect Avenue, PO Box 1, Osceola, WI 54020-0001 USA.

Library of Congress Cataloging-in-Publication Data
Matras, John.
  Mazda RX-7 color histor y/John Matras.
     p. cm.—(MBI Publishing Company sports car color history)
  Includes bibliographical references (p. 127) and index.
  ISBN 0-87938-938-9
  1. Mazda RX-7 automobile—Pictorial works I. Title. II. Series: Sports car color history.
TL215.M39M373   1994
629.222'2—dc20   94-32040

On the front cover: First and third generation RX-7s. In the background is Dawn and Tim Eilber's 1985 GSL-SE, and in front is a 1994 R2 owned by Mazda Motor of America.

On the frontispiece: The Infini was a limited production, high performance version of the 1991 RX-7. Minimal in its comfort accommodations, the stripped down Infini dashed from 0–60mph in just 7.0sec. Unfortunately it was unavailable stateside, or anywhere but Japan for that matter.

On the title page: The 1980 RX-7 GS was beloved among the enthusiast press with *Car and Driver* declaring, "You'd gladly trade your favorite fantasy for an RX-7 after one quick test drive."

On the back cover: The GTU[S] version of the second generation RX-7 was a one-year-only offering for 1989. *(inset):* Rotary engines require little space for the performance they offer. This 13B powered the 1987 RX-7.

Printed in Hong Kong

Contents

	Acknowledgments	6
	Introduction	7
Chapter 1	**Rotary Roots**	9
Chapter 2	**Fishing Boats to Rotary Rockets**	17
Chapter 3	**Mazda 1967 to 1978** *The Rotary Roller Coaster*	25
Chapter 4	**Project X605** *Gestation of Genius*	31
Chapter 5	**Evolution** *The 1979 and 1980 RX-7*	39
Chapter 6	**1981** *RX-7, Refined and Defined*	45
Chapter 7	**1982 and 1983** *Detail Improvements*	51
Chapter 8	**1984–1985** *The GSL-SE and Other Changes*	61
Chapter 9	**1986** *Finally, a New RX-7*	69
Chapter 10	**RX-7 Competition 1986–1988** *A New Era*	81
Chapter 11	**1988** *Diversification*	85
Chapter 12	**1989–1992** *The Major Minor Change*	91
Chapter 13	**RX-7 Competition 1989–1993** *Less is More*	99
Chapter 14	**1993–1994 RX-7** *Leaner and Meaner*	107
Chapter 15	**The Future**	117
Appendices	**Specifications**	120
	Sales	126
	Bibliography	127
	Index	128

Acknowledgments

I'm not the first author to write this, but this book would have been much more difficult had it not been for the help I received along the way. I'd like to thank as many as I can.

Thanks go to Steve Potter for too much to mention here, as well as to Stephanie Coull, wherever you are now. I also have to thank Fred Aikins of the Mazda Information Bureau for, well, Mazda information and some of the Mazda photos. Thanks also to Mitch McCollough and Jack Pitney of Mazda Public Relations for the help they were able to provide. Thanks to Margot Katz (Borman Associates) for the Mazda/Bose slides.

Thanks to the following individuals, in no particular order, who shared their RX-7s for the color photographs in this book: Cynthia A. Sgroi (blue 1985, 28,000 miles with the original tires!), Pete Joseph (black 1988 convertible), Leo Schefer (yellow early-1979 RX-7), Karen E. Elwell (black 1980 RX-7 GS), Robert Seton-Harris (red 1987 RX-7 Turbo), Jason Breiner (1988 Tenth Anniversary Turbo), Dave Nugent (black 1981 RX-7 GS), John Younger (white 1983 RX-7 RS), Ronald Arndt (red 1990 GTU^S), Larry Godshalk (blue 1987 Sport), Darren Graham (1979 RX-7 S details), Pete Giatrakis (gold 1986 RX-7), Terry Warren (1982 RX-7 GSL details), Gary Rebholz (red 1991 RX-7 Coupe), Pete Ernst (blue 1980 RX-7 GS), Frank Kresge (copper 1983 RX-7 GS with steel wheels), Craig W. Sayre (white 1984 RX-7 GSL-SE), Ralph Vecchio/Colonial Auto Sales, Bartonsville, PA (red 1988 SE), Charles W.L. Shefer (blue 1985 RX-7 GSL-SE), and Dawn Eilber (silver 1985 GSL-SE).

I'd also like to thank Otis Meyer, *Road & Track* librarian extraordinaire, for unearthing photos (and thanks as well for his patience); Jonathan Stein for allowing me to search *Automobile Quarterly's* files; Martin Peters at BF Goodrich; Mark Leddy at Chevrolet; Jim Tanner at Racing Beat; Al Dooley at Pacific Avatar; and Richard Dole and Tim Cline, for photo help. Thanks also to Charles Krasner, John Peter, and Marilee Bowles at *AutoWeek* for additional photo help.

Thanks also go to Gary Lowe and Steve Moorhead of the SCCA Northeast Pennsylvania Region; Karen Elwell and Dave Nugent of the Mazda Sports Car Club of Washington (P.O. Box 7212, McLean, VA 22106-7212). Thanks, too, to John Heilig at *Automobile Quarterly* and Ron Sessions of *Road & Track Specials*. And a thank you to W.C. Campbell III for logistical support.

Thanks to Zack Miller and Michael Dregni at Motorbooks International for turning the manuscript and the pile of photos into a book.

And finally, thanks to my daughters Katie, Cari, and Mandy for their enthusiasm and, last, but in no way least, to my wife Mary Ann for her aid and support and for generally putting up with me.

Introduction

Mazda's ads were unabashed in their prediction of classic status for the new RX-7. Not that advertising types are ever bashful or modest. On the other hand, it takes a lot of nerve to hold your product alongside true classics and demand a comparison.

Fortunately, the new RX-7 was up to the challenge. The press reaction was enthusiastic, and, more importantly, the car buyer reaction was sufficiently positive to make Mazda Motors of America revise its sales estimates upwards enough to keep Toyo Kogyo, as the RX-7's maker was then known, on a climb to profitability. Additionally, the RX-7's popularity absolutely guaranteed the survival of the rotary engine, which after having been "the engine of the future," had been marginalized at best and moribund at worst.

"In 1947, it was the MG-TC. In 1953, it was the Corvette, In 1970, it was the 240-Z. This year it's the Mazda RX-7. Now it's your turn."

The RX-7 is now into its third generation and is by every account a world class sports car, as good as anything from any carmaker in the world. It has won titles in virtually every form of auto racing for which it is eligible. In some classes in some years, the only debate was about which RX-7 driver would win.

The RX-7 is indeed a classic.

This book traces the history of the RX-7. Not only are the facts and figures provided, but information about each generation's development is presented as well. The book also covers the history of the rotary engine and other cars powered by it—Mazda wasn't the only maker of production rotary-powered cars—and those cars that almost were. And finally, for some really risky business, we will take a brief look into the future.

It's been a heck of a ride. Whether you were there the first time around or if this is your first exposure, I hope this book captures and replays some of the excitement of the Mazda RX-7. Enjoy the RX-7. As another 1978 Mazda advertisement said, "If not now, when?"

Chapter 1

Rotary Roots

On February 1, 1957, Dr. Walter Froede fired up the first Wankel engine, setting in motion not only a deceptively complicated contrivance but a chain of events that would inevitably lead to the Mazda RX-7. The little test engine proved the rotary concept, but it was still only the embodiment of an idea. Much work would be required and many years would pass before Mazda would develop Felix Wankel's theory into its ultimate expression.

The changes required to make Wankel's engine practical would at first infuriate him—though he would later express gratitude for Mazda's perseverance. He was, undoubtedly, a genius, one of those rare individuals with the talent to create anew. However, the idea of a rotary engine was not in itself new, preceding Wankel's birth on August 13, 1902, by centuries. In fact, one of Wankel's projects had been to catalog the various rotary engine concepts and ideas.

It is remarkable that Felix Wankel invented anything mechanical. World War I left him fatherless at age twelve, and the staggering inflation of the German Mark after the war decimated his inheritance. Wankel went to neither technical school nor the university but to work for a Heidelberg bookseller in 1921. An injury restricted him to the storeroom, but there he was undisturbed as he worked at his technical drawings and night and correspondence school work.

It is remarkable that Felix Wankel invented anything mechanical.

Though already enchanted by the idea of a rotary engine, he made his living devising sealing methods for the reciprocating piston engine, most notably by using gas pressure rather than strong springs. Wankel even achieved an engineering first by sealing a rectangle. His prestige was such that when invited in 1936 to join the Berlin-based Central Research Establishment of the Reich's Aviation Ministry, he was allowed to keep his own workshop in Lindau.

In December 1951, Felix Wankel came to motorcycle builder NSU (*Neckarsulmer Strickmachinen Union*) to develop a motorcycle engine with a rotary disc valve. Still he harbored the dream of a rotary engine. Threatening to take his talents elsewhere, Wankel in 1954 coerced an agreement from NSU to share the costs of rotary engine development.

The initial design was christened DKM

The thoroughly modern NSU Ro80 remained on the market from 1967 through spring 1977.

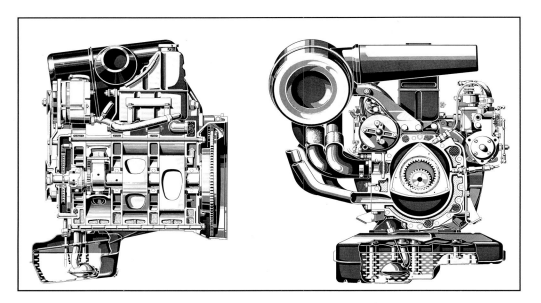

Mercedes-Benz cured low-speed operation problems with direct fuel injection, visible in this end cutaway view of a C111 three-rotor engine. Automobile Quarterly Photo and Research Library

53, for *Drehkolbenmotor* (German for "rotary piston motor") and the year Wankel started work on it. He variously attributed his idea, a three-sided convex rotor spinning in a peanut-shaped housing, to a liverish state from having eaten too much 1953 Christmas pudding, or to an idea he had had since the 1920s.

Although elegant in engineering terms, the DKM 53 and the subsequent DKM 54 were not practical. Wankel's designs not only had a rotating inner rotor, but also a revolving outer housing as well. The spark plugs were located on the rotor faces, while the intake and exhaust ports were located on the housing. The design required a second housing to hold the working housing from which the power impulses were taken. The pure rotary motion made the engine almost perfectly smooth, but getting the fuel/air mix in and the exhaust out was a nightmare, and spark plug replacement required complete engine disassembly.

No wonder that NSU's Dr. Walter Froede opted for a simpler design, secretly working around Wankel who had earlier dismissed it. Froede's experiments involved an engine with a stationary housing with a rotor eccentrically mounted on a central output shaft. But this was planetary, not pure rotary, motion and therefore displeased Wankel.

Wankel was more than displeased when Froede's engine, designated KKM for *Kreiskolbenmotor* (or "circuitous piston engine"), first ran on July 1, 1958. He called the KKM a "cart horse" compared to his "race horse," but the fate of the DKM was sealed. The KKM was not as smooth, nor able to run at the DKM's power-generating 25,000rpm, and the KKM's loopy orbit complicated sealing, but its overall simplicity was undeniable.

NSU, though, had been in dire straits financially, as Germans switched from cycles to small cars as the economy improved. NSU's own small car, the Prinz, would not appear until 1958. Relief seemed possible only if NSU could interest a major auto company in the rotary engine.

Interest came, but not from a car company. In May 1958, NSU gave Curtiss-Wright, an American aircraft engine manufacturer, one of Froede's engines to examine. After a weekend sequestered in his workshop with the engine, Max Bentele, the New Jersey-based firm's chief engineer, said that, despite problems, the engine was worth investigating. NSU derisively declined an offer of $50,000, then accepted $2 million, with $750,000 up front, for an exclusive license for the United States.

Curtiss-Wright set up a secret development operation: at first only Bentele, Charles Jones (Curtiss-Wright's stress and applied mechanics engineering chief), and a secretary were allowed into the "White House" where experiments and testing were performed. But when Bentele estimated that eight years would be required to develop a marketable automotive engine, Curtiss-Wright president Roy Hurley drafted forty-five of his best engineers to the project and engines of various sizes and configurations were produced, from small (4.3ci) air-cooled units for lawn mowers and chain saws, to a giant 1,920ci rotary that stood as tall as a man but had a combustion chamber too voluminous for complete combustion. It ran, however, making 800hp. The workhorses, however, were the single-rotor RC1-60, a 60ci engine, and the twice-as-large twin-rotor RC2-60.

The Comotor twin-rotor engine was available only with a three-speed automatic in the GS Birotor. Automobile Quarterly Photo and Research Library

The experimental NSU KKM 400 single-rotor engine became the first rotary engine to power an automobile when tested in NSU's Sport Prinz. Automobile Quarterly Photo and Research Library

The RC2-60 was installed in several Detroit chassis, including a 1965 Mustang that was road-tested in several magazines. Ultimately, Curtiss-Wright never produced automotive rotary engines but made money for a while selling sub-licenses to General Motors, Ford, and others, as well as collecting royalties from Toyo Kogyo for every Mazda rotary sold in the United States.

The Curtiss-Wright connection provided the necessary cash and publicity for NSU to keep its own rotary research and development on track. *Auto Motor und Sport* reported a mysterious Prinz fitted with a rotary engine in 1960, while NSU debuted the world's first rotary car at the Frankfurt Auto Show in October 1963. Henry Manney, writing in *Road & Track*, called the roadster "a Show Sensation."

The NSU Spyder at the 1963 Frankfurt show was very close to production and was offered for sale in 1964. Based on the Prinz chassis, the Spyder had a 498cc single-rotor engine hidden beneath the floor of a shallow rear trunk. A single two-barrel sidedraft Solex fed a peripheral intake port; the exhaust port was peripheral as well. It looked, said one observer, like an Alfa Spider, only smaller. With an 80.3in wheelbase, it was only 140.8in front to rear.

Although total NSU car sales were 5,000 per month, the Spyder sold only about 1,000 the first year. Yet that was an achievement, considering the price was DM 8,500 (about $2,000 U.S.), which was expensive, especially in Europe, for a revolutionary and unproven concept.

The NSU distributor for the United States was Trans-Continental Motors, run by Fred Oppenheimer. A *Road & Track* road test regarded the $2,998 Spyder more as a phenomenon than an automobile. It was called "peppy" above 3000rpm yet took 17.4 seconds to go 0–60 mph. Not bad for a 30.4ci engine, but the magazine "look[ed] forward to testing a larger, multiple-chamber version more suited to the American taste for torque."

Putting the Spyder on the road was an experiment for NSU, and its customers/testers did indeed find problems, typically in cooling and ignition, but especially in the seals. The carbon apex seals at the rotor tips became brittle and then broke when subjected to urban driving, and leaking oil seals on

The Citroen GS Birotor was doomed by the double whammy of high initial purchase price and excessive fuel consumption. Automobile Quarterly Photo and Research Library

rotor sides made the little engines smoke madly. NSU, however, provided free replacement engines for those which couldn't be repaired. Said Oppenheimer later, "I had a Wankel cemetery in New York for NSU."

The Spyder may have been something of a toy, but NSU's next rotary targeted the very serious Mercedes/BMW class. Entering production the day it was introduced at the 1967 Frankfurt Auto Show, the NSU Ro80 was the "ideal" car. The product of NSU veteran Ewald Praxl, the Ro80 not only had a 115bhp twin-rotor 996cc engine driving the front wheels, but also an aerodynamic profile, with a low rounded hood made possible by the small engine. The transmission was a three-speed "semi-automatic" equipped with a torque converter to damp the peripheral port engine's idle roughness, and an electrically controlled clutch that disengaged when the gearshift was touched for shifting. Suspension was fully independent, and the front brakes were mounted inboard.

Although it was received, as one historian noted, "with rapture" and won several "car of the year" awards, problems began almost immediately. An *Auto Motor und Sport* survey revealed that half of the engines were replaced under warranty, and passing Ro80 drivers had a special salute—holding up fingers for the number of apex seals replaced, the average reportedly three.

Although stronger alloys were found that greatly reduced apex seal failures, it was too late to save either the Ro80 or NSU's independence. In 1969, NSU was merged with Auto Union, a subsidiary of Volkswagen, and the K-70, a front-drive four-cylinder sedan developed by NSU appeared with a VW badge. A new corporation, Audi NSU Auto Union became a wholly owned subsidiary of Volkswagen in 1971. The Ro80 dribbled out of the factory through the spring of 1977, during which time a total of 37,400 were built. A replacement rotary sedan under development wasn't produced—the bloom well off the rotary rose by then.

During the 1970s, NSU wasn't alone in selling rotary-engined cars in Europe. Citroen began experimentation by putting single-rotor NSU-built engines in versions of its Ami 6, offering 500 to selected customers for $2,700 each. The customers, Citroen owners all and all who agreed to drive 20,000 miles per year, were to follow care rules that included inspections by Citroen's research and development staff. Called the M35, the coupes had a shape only a Frenchman could love or even tolerate. Citroen published a "legal" engine displacement of 955cc (nominal displacement was half that), and the carbureted engine produced a mere 50.6lb-ft of torque at 2745rpm and 54hp SAE at 5500rpm. A contemporary European tester noted that "the M35 could be driven properly only by shifting gears all the time—whether the car was slowed down in traffic or the

Felix Wankel, inventor of the trochoidal rotary engine, holds a DKM rotor. Note the spark plug holes in its faces and the ports in its sides. Automobile Quarterly Photo and Research Library

road was slightly uphill."

Citroen had already joined NSU to establish Comotor in 1967, a cooperative venture to build rotary engines for use by both companies. But when the engine was ready, only Citroen had a car for it, the new GS. An award winner with a piston engine, the GS was called the Birotor with the twin-rotor peripheral port engine. An automatic was mandatory. Alas, the OPEC oil crisis was in full swing in 1974, and the French rotary proved thirsty. With a purchase price double the francs of its piston-engined counterpart, the GS Birotor was short-lived and limited in number.

Daimler-Benz also showed interest in the rotor-motor growing up in its backyard. The Wankel-engined Mercedes-Benz C111 appeared in 1969 causing considerable speculation that a mid-engined GT version would go into production. Three-rotor and four-rotor versions were developed. The three rotor, with each chamber displacing 600cc, made 320hp. The Daimler-Benz rotary had peripheral intake ports, and direct fuel injection with two Bosch mechanical injectors per chamber. The 405hp four-rotor C111 had a top speed of over 170mph and did 0–100mph in the 14sec range. Millionaires were all but stacked up at Stuttgart's door, blank checks in hand; however, the C111 did not become a production model but was further developed as an advanced diesel engine test bed.

Nissan took a Wankel license in 1970 as the rotary potential was emerging and was able to display a prototype engine, a peripheral-port, dual-plug twin-rotor design, at the Tokyo Motor Show in the fall of 1972. A sports car to bow in early-1974, and based on the Sunny/210 chassis, was to be the apparent rotary recipient. Production was forecast at 10,000 cars per month. Nissan was serious: The license cost several million dollars and covered engines from 30 to 230hp with production anywhere in the world. But the OPEC oil embargo spiked the rotary Datsun. In mid-1974, Nissan announced that its rotary-powered model, once delayed, was permanently canceled.

Toyota didn't go in nearly so deeply for the rotary. Toyota acquired licenses from NSU and Wankel in 1971 (at the same price Nissan paid) but then negotiated to buy rights to Toyo Kogyo's patents and then entire engines, with rumors rife that Toyota wanted to buy the smaller company outright. Negotiations continued until the fuel crisis brought them to a halt.

In the United States, Chrysler and Ford had publicly negative attitude, toward rotary engines. Most likely, Chrysler was not in a financial condition to invest heavily in what was still a speculative venture, though it did develop a rotary air conditioner compressor, using the Wankel configuration as a pump. Although tooling was ordered, the compressor never went into production.

Ford Motor Company couldn't afford to take number three Chrysler's wait-and-see stance. Although Henry Ford II openly belittled the rotary engine and probably believed most of what he said, when General Motors embraced the rotary as the engine of the future, Ford couldn't afford to be caught out if its rival's predictions proved correct. First attempts to acquire technology or engines from Toyo Kogyo were rebuffed; Toyo Kogyo president Kohei Matsuda offered Ford a twenty-five percent share of the company *less the rotary engine* or, as one Ford insider was quoted, "twenty-five percent of nothing." Ford subsequently acquired—for $2.3 million—rotary rights through Ford-Werke, its German subsidiary. Under that scenario Ford needed permission from Curtiss-Wright even to sell rotary-powered cars in the United States. But an unpublicized provision would allow, after a payment of up to $20 million to Curtiss-Wright and about $30 million to NSU-Wankel, full worldwide manufacturing rights. Behind the scenes Ford worked on various rotary designs, but pulled the plug in mid-1974.

The biggest player in the rotary game had to be General Motors, at least if budgets were considered. Indeed, the General prepared impressive prototypes and came very, very close to putting a rotary-powered car into production. The first public showing of GM's efforts was at the 1973 Frankfurt Auto Show. The Corvette

The Corvette two-rotor debuted at the 1973 Frankfurt Auto Show. Automobile Quarterly Photo and Research Library

2-Rotor, as it was officially if not imaginatively named, was a show car but fully functional. Much smaller than contemporary Corvettes, on a wheelbase of 90in rather than 98in, it had an all-steel body—a first for any Corvette—but weighed only 2,600lb. The handsome sports car's shape was the product of GM design staff but was built by Pininfarina (who usually doesn't convert other people's designs into metal, though the General's money was probably persuasive). MacPherson strut front suspension, rack and pinion steering, and trailing arm rear suspension were all touted as firsts for a United States GM car.

The real news, however, was the two-rotor motor amidships. Designated as the RC2-266, the engine had two 1099cc chambers, which was calculated to be equivalent to 4.4-liter, or 266ci—hence the designation. The engine conformed to standard GM rotary-engine practice of the time, with cast-iron side plates and aluminum chamber housings, side intake ports, and twin ignition. Showing the seriousness of the project was the fitting of an air conditioner compressor! The Corvette 2-Rotor was mounted transversely inline with a GM three-speed automatic, the drive then returned to a central differential.

The 2-Rotor was joined several months later at the Paris show by the Corvette 4-Rotor. Mid-engined as well, it was actually the mid-engined 'Vette from the 1970 New York show but partially re-skinned and outfitted with a four-rotor GMRE (General Motors Rotary Engine) where the V-8 had been. The four-rotor engine was actually two two-rotor engines bolted together at a central housing from which gears drove the accessories and a Turbo-hydramatic transaxle.

Exciting as the rotary Corvettes were, they were still only show cars and merely smoke for the fire back in Michigan. Smoke was, in fact, reported to be one of the first impressions which the GM engineers had when they examined the NSU rotary engine in 1962 at the bidding of Ed Cole, then GM vice-president in charge of car and truck operations. But in a study, finished in late 1967, Ed Cole's son David suggested that the rotary's high exhaust temperature would in fact be beneficial with either thermal or catalytic converters. And reportedly, it was the fear of pollution control regulations that kept rotary prospects alive at GM and the rumor mill grinding, including one story that GM planned to buy Wankel G.m.B.h., the holding company for rotary patent rights, outright through Opel, its German subsidiary. (Just think of the antitrust prospects!)

Finally on November 10, 1970, NSU and GM signed a license agreement for nonexclusive manufacturing rights for the Wankel rotary, worldwide sales, and for any use except aircraft. As the GMRE project advanced in General Motor's Central Engineering Development Department, the decision was made that not a car division but rather GM's Hydra-Matic Division would manufacture the rotary.

General Motors was extremely tight-lipped about the GMRE project, spurring speculation, prognostication, and wild rumors from Wankel watchers. The truth was that General Motors was toiling very hard to make the GMRE work and return acceptable fuel mileage. The attraction wasn't rotary-engine advantages, which weren't sufficient to justify the immense retooling cost, but the ability of the rotary-engined Mazdas to meet—as early as 1972—the pending 1975 exhaust emissions rules.

The basic Wankel rotary engine yields a higher level of hydrocarbon (HC) and carbon monoxide (CO) emissions than the basic piston engine, but only about a third the oxides of nitrogen (NO_x). It was difficult at that time, however, to control NO_x in a piston engine without increasing HC and CO emissions. But because of the rotary's hot exhaust, HC and CO could be reduced in a thermal reactor—as Mazdas proved from the beginning of U.S. importation—and without increasing NO_x significantly.

With conventional engines, every manufacturer worried about meeting the 1975 emissions limits which threatened to shut down the entire automotive industry. Or so the manufacturers told the government while making pleas for delaying implementation of the lower limits. They were also telling the government other things behind closed doors, as the Environmental Protection Agency let slip while rejecting automaker pleas: "General Motors is the only major manufacturer which plans to produce limited numbers of rotary-engine vehicles in 1975."

How close the company came to that. Playing its cards close to its chest even with its suppliers—a typical source of information, General Motors was preparing an entirely new model for its GMRE. Though suppliers were told of a rotary-engined Vega, the new rotary-powered Chevrolet was in fact to be called the Monza, a sporty coupe without the public relations liabilities that had attached themselves to the Chevy econocar (though the name had previously been used for Chevrolet's ill-fated Corvair). The Monza not only had the styling to match its powerplant, but also a rotary-friendly chassis: a high driveshaft tunnel accommodated the rotary's output shaft, located higher in the block than in a piston engine.

However, when the new cars were introduced in the fall of 1974, the only thing the Monza lacked, in fact, was a rotary engine. Instead, the Monza came with either the Vega four or a 262.5cin V-8—except in California where it couldn't pass emissions tests. Variants appeared from Buick and Oldsmobile, the Skyhawk and Starfire respectively, with the V-6, which Buick had sold from 1962 to 1967, resurrected for these models. But plans for the GMRE were "uncertain."

It was Ed Cole himself, president of General Motors when the rotary engine license was obtained with NSU, who announced the delay. Only a week before

The Corvette four-rotor bowed at the 1973 Paris Auto Show. Automobile Quarterly Photo and Research Library

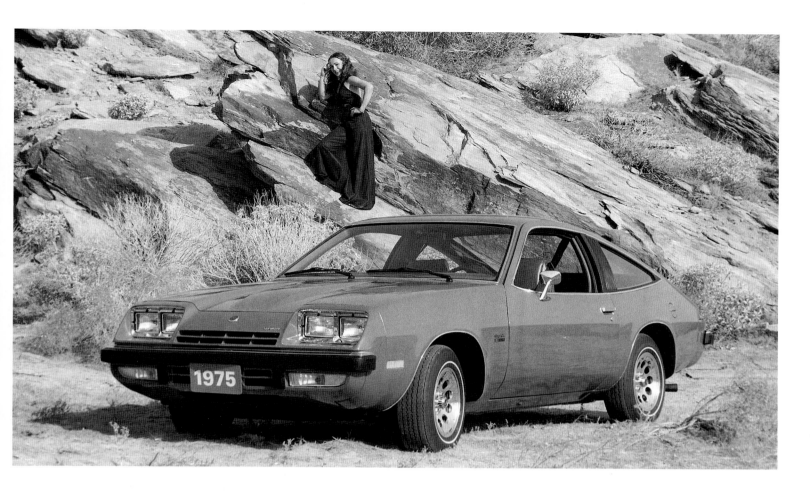

Plans for the new 1975 Chevrolet Monza to be powered by the GMRE rotary engine were scrubbed at the last minute. Chevrolet

his September 30, 1974, retirement, Cole commented, "Considerable progress has been made in the ongoing GM rotary-engine development program, but the lack of relief from the very stringent 1977 standards, which we are not sure can be achieved even with current production engines, makes it especially impractical to put into production any new engine which doesn't presently have the potential to meet those [HC and NO_x] standards."

David Cole summed it up even more succinctly: "They may think it's better to have one engine that can't meet [the 1977 emissions standard] than two that can't meet it." His retirement pending, the senior Cole himself indefinitely delayed the rotary.

General Motors' abandonment of the rotary engine had repercussions beyond Detroit. Immediately affected was AMC's Pacer. With the GMRE rotary not available for sharing, the unusual compact from American Motors was limited to using the venerable inline six, for which the Pacer wasn't ideally suited and which didn't suit the car-of-tomorrow image AMC had wanted.

The long-term consequence of the GM Wankel withdrawal, however, was the marginalization of the rotary. In 1972 it had been the automotive industry's golden child. Fichtel & Sachs of Germany made a line of rotary engines that powered everything from German Hercules motorcycles to American Arctic Cat snowmobiles. Yanmar Diesel of Japan had rotaries of 220 and 450cc powering outboard motorboats, riding lawn mowers, and snowmobiles. And Outboard Marine Corporation (OMC) developed a 35hp twin-rotor outboard motor displacing 1056cc and using side and peripheral porting, switching from side to peripheral at 3000rpm.

Suzuki exported a rotary-powered motorcycle to the United States market, but it was not a sales success, most likely because its rotary engine—and thus the cycle—was expensive and down on power because of a displacement-restricting license from NSU/Wankel. However, Van-Veen, a small Dutch cycle manufacturer, used a Comotor rotary to create a rotary superbike.

The smallest rotary was a palm-sized model aircraft engine made by Ogawa Seiki (OS) of Japan. Displacing a mere 5cc, the OS not only had novelty as an advantage over its piston counterparts, but the lack of vibration allowed radio equipment to last longer. The engines were expensive, however.

And finally, Rolls-Royce developed a "stacked" diesel rotary. A smaller "power" chamber was cast above a larger primary chamber that worked as a compressor. R-R planned no Rotary Ghosts, however—the engine was designed for use in military tanks!

In the end, Toyo Kogyo was left to struggle alone, selling its rotary-engined cars to a fickle public in a world of seemingly diminishing expectations.

Chapter 2

Fishing Boats to Rotary Rockets

The production Cosmo Sport differed only in detail from the show car of four years earlier.

"I was shocked," Kenichi Yamamoto told *Automotive News* in 1993 speaking of his appointment to head up Mazda's development of the Wankel engine in 1963.

"I had a rather negative view of the rotary engine," Yamamoto continued, "and . . . Japan was being rapidly motorized. The Japanese were anxious to have any engine, so I thought, 'Why should I take up the rotary engine?'"

It was a rather dull prize, in Yamamoto's estimation, for the development of the new Familia, which would form Mazda's midsize (for Japan) line, and for a career with Mazda that dated back to 1946. But the assignment came from then-president Tsuneji Matsuda, so Yamamoto threw himself into the project. The problem, Matsuda explained to Yamamoto, was that Japan's Ministry of International Trade and Industry (MITI) intended to merge Toyo Kogyo out of existence as a part of a general plan to consolidate the country's automotive industry. Matsuda believed that by adopting the rotary, Toyo Kogyo would develop a "technological charter" and a reason for maintaining its autonomy. "Rather than profit, we went after an identity and independence," Yamamoto told *Automotive News*.

> "Other companies considered the rotary a source of revenue, and if there was no profit, they gave up. We stuck to it with persistence."
> —Mazda's Kenichi Yamamoto

The ploy was successful. Toyo Kogyo, which had been manufacturing automobiles only since 1960, remained a separate company. The rotary engine made — and almost unmade—Toyo Kogyo. And it also propelled Yamamoto to fame well beyond that of most automotive engineers, indeed to the presidency and chairmanship of the company, by then renamed Mazda. And as Yamamoto himself would concede upon his retirement, "I am Mr. Mazda."

The word "Mazda" itself was rooted in the ancient religion of Zoroastrianism, the god of light. General Electric had used the name on light bulbs in the United States. To understand why a Japanese company chose a name from a Persian religion it is necessary to go back to August 6, 1875,

17

Mazda's Founder: Jujiro Matsuda ultimately forsook fishing to lead what would become one of the world's most innovative automobile companies. Automobile Quarterly Photo and Research Library

and the birth of Jujiro Matsuda, the twelfth son of a poor Japanese fisherman. When Jujiro was three, his father died leaving Jujiro to spend his early childhood fishing with his brothers. But he abandoned the sea at thirteen to become an assistant in a small blacksmith shop. It clearly suited him, for by age nineteen Matsuda had a metalworking shop employing some fifty workers. That business failed and its successor, the Matsuda Pump Partnership Company, did too. The next two businesses were successful, however, the latter bought by the Nihon Steel Manufacturing Company.

By that time, World War I was in full swing, depriving Japan of European cork. A domestic substitute was the thick bark of the abemaki tree and a company called Toyo Cork Kogyo formed to bring it to market. But real cork returned at war's end, displacing abemaki cork and plunging Toyo Cork Kogyo into financial difficulty. The company was reorganized by its primary bank (then as now an important factor in Japanese industry), and Jujiro Matsuda was asked to sit on the company's board of directors. Shortly thereafter he replaced the company's illness-stricken president.

Matsuda changed the company's course, leading it out of agriculture and into manufacturing, and dropped "Cork"

Toyo Kogyo's Hiroshima factory as it appeared circa 1930 when vehicle manufacture was about to begin. Automobile Quarterly Photo and Research Library

from the company name in 1927. Thus began Toyo ("orient") Kogyo ("industry") Kaisha ("company"), Limited.

Much of the company's work was for Japan's growing military. It was lucrative but fluctuated with the political ebb and flow. Matsuda saw more potential in civilian products. A test run of thirty motorcycles in 1930 led to the manufacture of three-wheel-trucks, essentially cycles with two rear wheels. Small but capable of carrying up to two tons and turning almost in their own length, the trucks were ideal for the narrow streets and alleys of Japan's industrial districts. Toyo Kogyo became a leading truck manufacturer, adopting "Mazda" as a trade name. Why "Mazda" is uncertain, but it does sound remarkably like Matsuda when spoken by the Japanese.

In the 1930s Japan's automobile industry was small and mostly served the commercial market. In 1936, Japan built 5,004 trucks and buses, 6,335 small vehicles including three-wheeled trucks, but only 847 cars. Nonetheless Toyo Kogyo displayed a prototype automobile in 1940. It proved for naught when the company's resources turned to war production for the next five years. Amazingly Toyo Kogyo's Hiroshima factory and offices were spared major damage during the war. Even the atomic bomb left the plant standing.

With the hostilities past, Mazda resumed commercial production of three-wheel-trucks. A young Kenichi Yamamoto worked on the assembly line. The Hiroshima native received an engineering degree from the elite Imperial University (later Tokyo University) in 1944 and was immediately conscripted to supervise construction of kamikaze aircraft. After the war, Yamamoto, then about twenty-three, threw himself into bolting together Mazda truck transmissions, the only job he could find. "I worked like a workaholic," he recalls. His technical talents were eventually noted, however, and Yamamoto was transferred to the Engine Design Department where he designed an 1157cc V-twin for the 1950 Type CT three-wheeled-truck.

Other Japanese manufacturers built cars under license: Nissan an Austin, Isuzu a Hillman, and Hino a Renault. Mitsubishi built Kaiser-Fraiser Jeeps from kits. Toyota and truck-maker Mazda were unusual for building their own designs.

But Japan was in what *Time* magazine

Thirty prototype motorcycles built in 1930 but not put into production. Automobile Quarterly Photo and Research Library

called an "orgy of expansion." Yamamoto was not the only workaholic in Japan, and investment was actively supported by the government via special tax allowances. Economic growth was typically in double digits and from 1955 to 1962 car building saw a tenfold increase.

Toyo Kogyo was the undisputed king of three-wheeled trucks. Counting these alleyway workhorses, the Hiroshima firm was Japan's premier motor vehicle maker. Toyo Kogyo trailed only Nissan and Toyota in conventional vehicles. Yet the three-wheeled truck was doomed, peak-

By 1960, Toyo Kogyo would become Japan's number three vehicle builder based on these three-wheeled trucks. Automobile Quarterly Photo and Research Library

ing in 1960, the year that the first Mazda car was introduced. The Mazda R-360 was an undeniably cute two-door coupe propelled by a rear-mounted air-cooled 356cc V-twin. As a sub-360cc (*kei*, or "light" car), it had low registration fees and owners in Tokyo weren't required to show proof of a parking place.

It was just the beginning. The 1962 two-and four-door P-360 Carol captured two-thirds of the sales in its class. The Familia, with a front-mounted 782cc four cylinder driving the rear wheels, debuted in 1964. Its development was complete in May 1963 when Yamamoto was transferred to the rotary engine project.

Tsuneji Matsuda, succeeding his father in 1951, first learned of the Wankel rotary engine in 1959 from Curtiss-Wright. The corporate psyche was piqued and the company requested a license in early 1960. NSU reacted coolly toward Mazda and to the Japanese industry in general. Still, Matsuda and a contingent from Toyo Kogyo visited NSU in October 1960. They were amazed by NSU engines on test stands and coins balanced on smoothly running engines, and impressed by NSU Prinz test cars with 25,000 miles of testing. Toyo Kogyo was granted a license, for sales in Japan only, on October 12, but bureaucratic delays in Japan de-

Prototype passenger cars were built in 1940, but the war halted their entry into production. Mazda

The 1960 Mazda R360 was the company's first production automobile. Mazda

layed approval until July 4, 1961—after Yanmar Diesel.

Led by Executive Vice President Kohei Matsuda, Toyo Kogyo visited NSU again in July 1961 to ask why the NSU rotary wasn't in production. They returned with drawings and technical information and promises for an NSU KKM 400 to be shipped to Hiroshima. But before its November 1 arrival Toyo Kogyo had built its own rotary from the drawings. Toyo Kogyo engineers were disappointed with the NSU design, citing excessive vibration at idle, clouds of white exhaust smoke, and appalling oil consumption. Worse, at about 200 hours, power output dropped suddenly, seal chatter having eaten the chamber electroplating.

After heading the design of the R360, Kenichi Yamamoto considered the assignment to develop the rotary engine to be a demotion. Automobile Quarterly Photo and Research Library

Other companies would have given up, and many others then did. But as Yamamoto told *Automotive News* some thirty years later, "We stuck to it with persistence. Why? Other companies considered the rotary a source of revenue and if there was no profit, they gave up. Mr. Matsuda showed leadership. He championed this unique engine. I gave my best effort because he was my champion, and I thought I should be the champion for my staff."

The early Mazda rotary engines had two spark plugs per chamber, and two completely independent ignition systems. Automobile Quarterly Photo and Research Library

From the rear, the Cosmo Sport had a slight T-bird look.

A $750,000 rotary engine development center was built, staffed by 180 employees and equipped with the latest computers for engine analysis and comparison. By August 1964 about thirty engine test benches were in operation. Yamamoto established a policy of evaluating the fledgling rotary by the standards applied to the well-established piston engine because that's what customers would do.

The Toyo Kogyo rotary almost immediately differed from NSU's. Carbon-aluminum apex seals replaced NSU's cast iron and Toyo Kogyo also went with dual, instead of single spark plugs, timed to fire in sequence. Toyo Kogyo also switched to side ports for intake, with one port on either side of the rotor. This significantly reduced "overlap" of intake and exhaust port openings, and though sacrificing power at speed, the improved idle smoothness was worth it. Finally, to guarantee smoothness, Toyo Kogyo worked only with twin rotor designs.

The special engine deserved a unique home, so work on a chassis began in December 1962. It was no *kei* car. Befitting its "engine of tomorrow," the car was almost all new and a sports car from one end to the other. The prototype was a Tokyo Motor Show debutante in late-1964, though Tsuneji Matsuda had actually driven one of two existing 798cc L8A-powered prototypes to the 1963 Tokyo show and back home to Hiroshima, stopping to visit Mazda dealers along the way.

Dubbed "Cosmo," the two-seater prototype had a pointed prow and faired headlamps. Its roof, a nondetachable hardtop, had a Ford Thunderbird look, though a burnished metal targa band was more reminiscent of the Crown Victoria. At the rear large "rocket exhaust" taillamps were split by a blade-type bumper. An attractive shape overall, it was troubled by excessive details: vents, ridges and such, not to mention garish full wheel covers! At least the interior looked right, with a three-spoke wood-rimmed steering wheel and a classic British sports car dash. Nor did Toyo Kogyo skimp on suspension, using double A-arms in front and a genuine deDion axle at the rear. The engine for the 1964 show car was a Type L10A displacing 982cc.

The car looked ready to go and it was. Only months later, in January 1965, sixty Cosmos were turned over to ordinary drivers and Mazda dealers and distributors across Japan, for real world testing. Different engine porting configurations provided additional data. The Cosmos tallied 375,000 test miles with no mechanical failures.

When production began in May 1967, the Cosmo was equipped with a dual side port engine and a four-barrel Hitachi carburetor. The twin rotor engine displaced 982cc and produced 108 or 110bhp, depending on where you read it; from the latter figure came the Cosmo's "110S" export designation. The Mazda rotary, with dual ignition, had the unusual feature of two distributors. The only transmission offered was a four-speed manual with direct drive on high and a final drive ratio of 4.111:1.

Front suspension, as on the show cars, was unequal length A-arms with coil springs, an anti-roll bar, and tube shocks. The upper arms were welded tubes, the lower arms stamped. The rear suspension featured a 51mm deDion axle on semi-elliptic springs, with trailing links and tube shocks. The differential was mounted to the body on a T-shaped member with noise and vibration insulated by rubber. Axle half-shafts used ball-bearing sliding splines. Stamped steel wheels measured 14x4.5in with either 6.45x14in bias ply or 165HR-14in radial Bridgestone tires.

Neither brakes nor the rack and pinion steering had or needed power assist; disc brakes were used on the front wheels with drums at the rear.

Production styling was tidied by elimination of the targa band and addition of cabin-ventilation "ears" (no air conditioning). Otherwise little changed from the original show/test cars. Over fourteen months, 343 of the L10A Cosmo Sports were built.

In July 1968, production switched to the Type L10B Cosmo. Although the designation refers to a new engine, the entire car was subtly though substantially

The interior had all the right sports car touches.

changed. The new engine still displaced 982cc but revised port timing boosted power to 128bhp at 7000rpm. Fifth gear (overdrive at 0.841:1) was added to the transmission, the brakes got vacuum assist, and air conditioning became optional. Fifteen inch wheels with 155HR-15in radials became standard equipment.

The Cosmo also grew, gaining 5.9in between axles, visible in more distance between the rear edge of the door and the leading edge of the rear wheel opening. Overall length, however, decreased slightly. From the front, a larger, mesh opening for the radiator could be seen below the front bumper which gained rub strips at the corners.

Performance was noticeably improved. Whereas Toyo Kogyo claimed 16.4sec for the quarter mile and a top speed of 115mph for the L10A, the L10B tripped through the quarter in 15.8sec with two up and went 124mph—a full 200km/h!

The price rose as well. The L10A was priced at 1.48 million yen, or about $4100 at the 1967 exchange rate. That wasn't cheap, but as Mazda partisans will point out, it was less than two-thirds the price of the Toyota 2000GT. The L10B Cosmo retailed at 1.58 million yen, or about $4390, still cheaper than Toyota's exotic.

Production of the Cosmo crept along—literally. The assembly line was not automated, instead resembling (as much as possible for the Japanese) that of an Italian exotic manufacturer. Cars moved along on chassis dollies pushed by hand. No wonder that only about twenty Cosmos per month—about one per work day—were completed. Toyo Kogyo built the Cosmo until September 1972. The final count of L10B Cosmos totaled 1,176.

The five-year production total of 1,519 Cosmos of both types makes the model seem less than a success. Factor in development cost and Toyo Kogyo surely lost money on every one. But that misses the point. The Cosmo was intended as a test bed for the rotary as well as a high visibility vehicle for the rotary engine. And that it was.

Toyo Kogyo brought the Cosmo 110S to market in an amazingly short time, particularly for a car with a new type of engine. Unfortunately the Cosmo had no direct successor. Later cars would use the name and concepts pioneered in the Cosmo but would get little further than Toyo Kogyo drawing boards—until, that is, the RX-7.

Chapter 3

Mazda 1967 to 1978
The Rotary Roller Coaster

The RX-3 went into production in time for an April 1972 debut in the United States. Mazda

While the Cosmo 110S would find an eventual heir in the RX-7, Yamamoto's rotary engine would quickly find a home in Yamamoto's Familia. Debuting at the 1967 Tokyo Motor Show as the RX85, the new model simply had a 10A rotary (the "L" dropped when the engine was detuned for production) where the four-cylinder piston engine had been. A smaller carburetor and milder port timing dropped power to 100bhp at 7000rpm.

The R100 was to be Toyo Kogyo's first volume production rotary-engined car. Two-seaters were officially frowned upon in Japan—and still are, unless they make lots of money like the Miata—so the Familia coupe with its 2+2 seating would be much more socially acceptable. Putting a rotary engine in an existing piston-engine chassis also made economic sense, achieving multiple use from a given design.

In 1970, Toyo Kogyo had no sales or-

> *The rotary frenzy was international by 1972, and if a carmaker didn't have plans to introduce a rotary-powered car, it was investigating it.*

ganization in the United States. It hadn't been making cars all that long, and even Toyota and Nissan had only recently had any real success in the American market. For a beachhead, Toyo Kogyo chose Seattle, Washington, an unlikely location except that Toyo Kogyo had test-marketed in British Columbia in 1968. So, an import corporation, Mazda Motors of America (N.W.), Inc., was established in Seattle's suburbs and a shipment of 380 cars dispatched.

In addition to the rotary-powered R100, the shipment included the 1200 (as the Familia with a 1.2-liter engine was named for American consumption) in sedan, coupe, and station wagon form, and the 1800, a 1.8-liter sedan and wagon originally styled by Bertone. By the end of the year, there were 2,305 Mazda cars on American roads.

Tsuneji Matsuda would not live to see it, however. Yamamoto's champion and the man he credits with Toyo Kogyo's adoption of Wankel's wonder, died in November. Leadership of the family-held firm was passed to his son, Kohei Matsuda.

American operations expanded to

California (Mazda Motors of America, Inc., or MMA) on May 1, 1971, and to Texas and Florida about the same time. The first shipment of 2,000 cars—seventy percent rotary-powered—arrived in California in mid-April and included the R100 and its piston-engined equivalent the 1200 coupe, and the new rotary-powered RX-2 coupe and sedan and its piston-engined equivalent, the 616.

The country was in a rotary frenzy and Mazdas were selling right off the transporter. One dealer reported using a bullhorn to control showroom traffic. With only a half-year's sales and dealerships located on the fringes of the country, Mazda had moved from nowhere the year before to thirteenth overall amongst import makes. Meanwhile, Toyo Kogyo produced its 200,000th rotary-engined automobile.

The rotary boom continued in 1972. The RX-3 debuted at the New York Auto Show on April 1 as a coupe, sedan, and the first rotary station wagon in the world. The RX-3 was smaller than the RX-2 and intended to replace the R100, which, based on the decade-old Familia, was becoming dated despite its rotary engine. The RX-3, however, had the same 12A engine as the larger RX-2 and so developed a reputation as a pocket-sized hot rod, especially the coupe. The RX-3's piston-engined equivalent was the 808, available in all three body styles. The R100 was officially still on the books, as was the 1800 station wagon, but the sedan had been dropped. The RX-2's piston equivalent got a 1.8-liter engine and was renamed the 618.

On December 1, 1972, midwestern dealerships opened, first in Chicago, with

The front-drive R130 used the 13A rotary, with different dimensions than the 12A/13B family of engines, and was available only in Japan beginning October 1969 for two model years. Mazda

The R100 was the first rotary-engined Mazda produced and exported in volume. Mazda

East Coast dealers added about the same time. It couldn't have looked better for Mazda. The rotary frenzy was international, and if a carmaker didn't have plans to introduce a rotary-powered car, it was investigating it. Mazda planned to be all rotary by 1975, and General Motors planned for 80 percent of its 1980 production to be rotary equipped. Only the rotary engine appeared capable of meeting the 1975 U.S. emission limits.

Feeling its oats, Toyo Kogyo launched the X020G project. The X020G was slated to go nose to nose with the Corvette and E-Type Jaguar; though just a 2+2, it was considered a large car by Japanese standards. Styling would be dramatic and suspension fully independent, with semi-trailing arms at the rear.

Its heart would have been a new engine. One proposal was a simple widening of the 12A, the equivalent of a bore job on a piston engine, resulting in a 1.5-liter rotary with a target horsepower of 135. The other was all new. Not only was this second engine wider, but the trochoid dimensions were larger as well. The 2.1-liter 21A would have produced 180bhp; a proposed 22A could have made an even 200bhp. Prototypes were built and run on test stands but would never go into a car. The X020G had reached the full-size clay model stage, but changes in the political and economic climate put a chill on rotary engine products worldwide.

How quickly times change. The Wankel's rise may have been *Automotive News*' story of 1972, but for 1973 the OPEC oil crisis was center stage. Mazda sales surged to fourth place among imports by April, and by June, Mazda had sold more cars than the entire previous year. However, by mid-June, gasoline shortages had the federal government imposing price ceilings on petroleum. Gas lines began to appear, and critics were calling for the abolition of the automobile.

About the same time, Mazda rotaries got a reputation for being fuel-thirsty. MMA general manager C. R. Brown cried foul, claiming that the cars not only met the Clean Air Act 1975 standards, but they also equaled "any current model domestic car offering equal performance."

Still, by December, Mazda was offering rebates on "hard-to-sell" RX-2 and RX-3 coupes and sedans to make room—physically—at the port for the incoming 1974 models. Ironically, however, Mazda was still ranked as high as fourth in the monthly import sales tallies, despite a run on other imports as Americans abandoned their big Detroit-built cars.

Still Toyo Kogyo persevered with the rotary engine. In fact, about ninety percent of Mazda's 1974 product would have a rotary under the hood. Among the incoming cars only the 808 coupe had a piston engine. A new rotary-powered car, the RX-4, was introduced as the marque's top-of-the-line, offered in the three basic versions: coupe, four-door sedan, and wagon. The

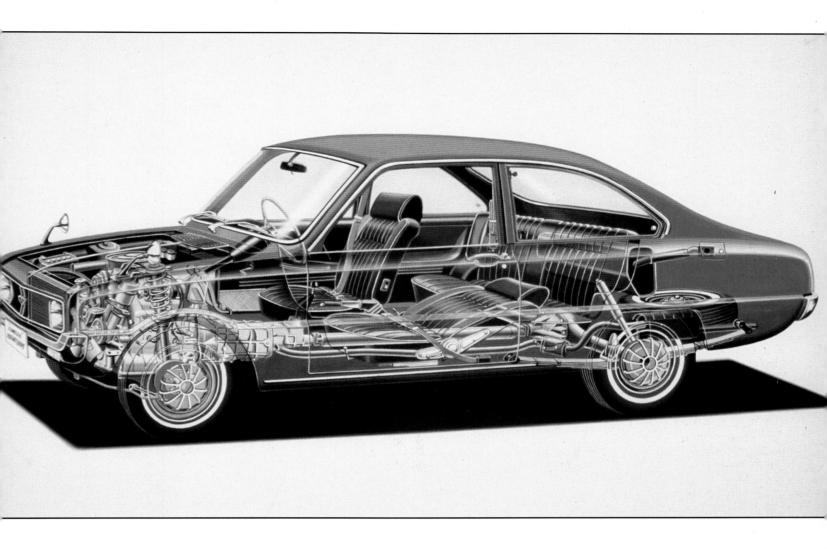

Rather than a hard-to-build sports car, Toyo Kogyo took the mass production expedient of installing the rotary in the Familia, calling it the R100. It became the first rotary-engined Mazda produced and exported in volume. Mazda

RX-4 had a new, larger engine, the 13B, made by widening the 12A by 10mm. The larger and more luxurious RX-4 could hardly have had better fuel economy, but the hope was that American consumers would compare the RX-4's fuel consumption to the traditional V-8. Another new model, the Rotary Truck, used the 13B in a specially modified Mazda minitruck, unavailable anywhere but North America.

Brown also badgered the Environmental Protection Agency to retest Mazda's fuel economy. EPA consented with a new test featuring two cycles, urban and highway, that showed the Mazda rotary comparing favorably with American V-8s. As a result of this test, the EPA changed its procedures for the 1975 model year, publishing city/highway mileage figures for all cars. It was a direct result of Brown's protest.

Unfortunately the damage was already done. The EPA results weren't published until April. Mazda sales for May 1974 were down by 30 percent, while Volkswagen and Toyota sales rose by half and Datsun sales doubled. To make matters worse, Mazda rotaries had earned a reputation for poor reliability: the side oil seals on the rotors of the early cars failed quickly, and there were problems with the apex seals. These were solved, with Mazda even boosting its rotary engine warranty to three years/50,000 miles. But questions (as well as a lawsuit) about how well some customers had been treated wouldn't be settled for years.

Brown hung tough, waiting for General Motors to introduce a rotary, saying "There will be fewer rocks thrown at us

Kohei Matsuda reigned over a turbulent era of growth and retrenchment but was an enthusiastic supporter of the RX-7. Automobile Quarterly Photo and Research Library

The Mazda Rotary Light Bus was a unique application of rotary technology, and it was the only rotary bus in the world. Automobile Quarterly Photo and Research Library

Pre-OPEC oil embargo: Posing with an RX-2, C.R. Brown, general manager, and Jiro Morikawa, president of Mazda Motors of America, still had something to smile about. Automobile Quarterly Photo and Research Library

once we are no longer alone with the Wankel." Yet by July, Brown would be gone. The forty-one-year-old, who had been MMA's first employee in Seattle, cited "personal reasons" for his leaving.

Regardless, his successor, Chrysler veteran Sidney Fogel, was unable to stop the slide. In September 1974, Fogel had no sooner told *Automotive News* that a lack of public confidence in the future of the rotary was Mazda's biggest problem, than General Motors announced another delay in introducing "the Wankel Vega." The resignation of GM president Ed Cole, one of the rotary's strongest advocates, indicated the postponement would be permanent. The Chevrolet Monza, especially designed for the rotary, would never be sold with anything but piston power.

Meanwhile blackberry vines were growing from under the hoods of Mazdas backlogged at the port. Rebates helped move once hot Mazdas in the U.S., while in Japan, Toyo Kogyo sent out assembly-line workers to sell cars.

The Mazda lineup at the start of 1975 included the 808 and RX-3 coupes and wagons; the RX-4 in coupe, sedan, and wagon form; and the rotary and piston-engined pickups. Sales recovered in Japan, but the final count in the U.S. was down about half. For fiscal 1975 (ending October 31), Toyo Kogyo suffered an operating loss of $44.4 million. Sumitomo Bank, Toyo Kogyo's principal financier, appointed its own executive director to the position of executive vice-president for the carmaker, diluting family control of the company. The change in "product" would be as significant: The total rotarization of Mazda automobiles would no longer be an eventuality.

Still, a new rotary-engined model was introduced in 1976. Called the RX-5 Cosmo in Japan, it was sold as the Cosmo in the U.S. A roaring success in Japan, it hit Stateside with a hollow thud. Based on the RX-4 chassis and powered by the 13B, it was a sort of Japanese Thunderbird, but it didn't appeal to American consumers, for whom the Cosmo had been expressly designed. In its road test of the Cosmo, *Road & Track* wistfully wished instead for a "new Mazda sports car built to compete with the Porsche 914 and Fiat X1/9, with either a single or double rotor . . ."

The other new model for 1976 was tagged with a real fuel crisis name: Mizer. This stripped 808 with a 1300cc piston engine had at least fuel economy as a reward; its 42mpg highway/32mpg city topped the EPA's 1976 mileage chart.

In late-1976 Kohei Matsuda told American journalists that a "sporty rotary engine coupe" would appear in 1978, as well as possibly a "sporty sedan." The sporty sedan appeared as the 1979 626, pleasant but with only an 80bhp four. That sporty rotary engine coupe would be the RX-7, though no one knew that yet. In the meantime, Mazda had to get through 1977.

Fortunately, the GLC—short for "great little car"—was the best news Mazda had received in a long time. The rear-drive hatchback had the same 1272cc engine as the Mizer, which it replaced, but it was a much more modern car overall. In the month after its introduction, it accounted for 54 percent of Mazda's sales, boosting total sales by 74 percent over the year be-

Although dedicated to the rotary, Mazda still produced kei-*class cars such as the Chantez, introduced in 1972, for domestic consumption.* Automobile Quarterly Photo and Research

fore for the same period. The GLC may have been humble, but it saved Mazda. A new rotary model was the RX-3SP, introduced in March, a striped and spoiler-equipped RX-3 coupe.

The only cars new for 1978 were additional versions of the GLC. Yet 1978 would be the end of an era for Mazda in the United States, as it would be the last year for the Cosmo, the last for the RX-4, and the last for the RX-3SP. There would be no new rotary sedans on American shores.

But it was by no means the end of the rotary engine nor rotary-engined Mazdas. There would be a successor to the RX-5, though for Japan only, and even today there's a three-rotor turbocharged Cosmo on the market. And better yet, in March 1978 (in Japan) and in April (in the United States), Toyo Kogyo would introduce the 1979 RX-7 and turn the first page of a whole new story, a story of rotary-powered sports cars.

The Rotary Pickup was sold only in the U.S. but only for a limited time. Road & Track

The 155mph Mazda RX500 experimental show car carried a 12A rotary midship. It debuted at the 1970 Tokyo Auto Show. Road & Track

Chapter 4

Project X605
Gestation of Genius

A preproduction final prototype undergoing testing in Japan. Mazda

There are many threads in a company's history and in that of its products, and nothing occurs without precedent. However, the true beginning of the RX-7 began in the depths of Mazda's 1975 financial crisis and the travels of two people, one a Mazda engineer named Akio Uchiyama, the other a Sumitomo banker turned Mazda board member, Sinpei Hanaoka.

From late-1969 Uchiyama had been project planner for RS-X, a rotary sports car proposition that got no further than engineering drawings and calculated performances. Similar in size to the Datsun 240Z, the two-seater—designated X020A—would have had a choice of three rotary engines: a 10A producing 82bhp, or two 12A versions with 97 or 170bhp. Ultimately, it simply served as part of the evolution of the ill-fated X020G, which fell to the oil crisis in October 1973.

Uchiyama was fated in 1975, along with other corporate employees of various duties and rank, to assist a Japanese Mazda dealer clear the backlog of cars that had accumulated in Japan as well as the United States. As luck would have it, his dealership assignment was in Tsu, near Suzuka, one of Japan's race tracks. By talking with spectators and amateur racers, Uchiyama learned that Mazda was wedded to the rotary in the public mind and came to the conclusion that recovery could only come through making the rotary work. Furthermore, the way to do that was with a genuine rotary-engined sports car.

The RX-7 was a mammoth hit. Road & Track *called it, "an enthusiast's dream come true."*

About the same time, director Hanaoka came to a similar conclusion but half a world apart. Having toured the United States in 1975 to learn more about marketing for that most important market, Hanoaka returned to Japan to recommend to the board that the company develop and sell in the United States a genuine rotary-engined sports car.

Project X605 began in the summer of 1976. Unlike a project that had superseded the X020G, X605 would not share a floor pan with a sedan (the earlier projects, the X516 sports car and X517 sedan made use of the same chassis by flipping curved front A-arms to alter the

Although the model year designation for the first RX-7s was 1979, production actually began in early-1978. Mazda

Note the two-spoke steering wheel and speedometer calibrated to 130mph on this 1979 RX-7 S.

wheelbase!). Toyo Kogyo president Kohei Matsuda is said to have taken a personal hand in the project, although Moriyuki Watanabe, deputy chief of research and development, was in charge of day-to-day operations (it served him well—he later became chairman of the board). Akio Uchiyama returned from his dealership assignment to work on chassis layout and assist Chief Project Engineer Sumio Mochizuki. Matasaburo Maeda headed styling, joined by exterior designer Yasuji Oda. At the time, Kenichi Yamamoto was responsible solely for rotary engine development.

Project X605 would be designed solely for rotary engine power. There would be no equivocating with piston-engined versions that would hinder the packaging advantages of the rotary; no production Mazda since the Cosmo Sport had used the compact size of the rotary as a design consideration. Although the 13B engine was available—it was still on sale in the United States in the RX-4 and Cosmo RX-5 when the RX-7 was introduced—Mazda opted for the smaller 12A for the RX-7. External size apparently wasn't a factor, as the 13B is only marginally longer than the 12A. Perhaps Mazda preferred the smaller engine's slightly better fuel economy over the few added horsepower of its larger sibling.

Another early decision was that the RX-7 would be front-engined. This had

Mazda designers wanted a one-piece backlight, but economics dictated a less expensive three-piece solution. Mazda

everything to do with the use of the rotary engine. Mid-engined designs, then rather in vogue, were utilized to center the bulky engine within a car's wheelbase, both to move its weight from above the front axle line and to decrease the polar moment, or the car's resistance to turning. The pocket-sized rotary, however, could fit behind the front axle line in a front-engined configuration, achieving a near 50/50 weight distribution. Mazda, in fact, didn't call the design "front engine," but "front mid-engined."

Toyo Kogyo's aim was to build a true sports car and to not repeat the errors of the RX-5 Cosmo. Construction would be unit body, but the chassis engineers achieved a remarkable torsional rigidity of 6,780lb-ft per degree, so good that it would be a goal for the second-generation RX-7 chassis.

Befitting a true sports car, the Project X605 would be small. Models sent to America would have no back seat, but because of official disapproval of two-seaters in Japan, a rear collision brace for U.S. market cars was not installed for domestic consumption so that vestigial rear seats could be fitted. Although bigger than the Cosmo 110S—about five inches longer and four inches wider—Project X605's dimensions placed it among contemporary sports cars. Only Fiat's X1/9 and 124 were lighter. The Mazda, however, had the longest wheelbase, longer even than the 280Z, whose overall length was five inches greater.

The Project X605 profile was decided fairly early as well, though detail variations were proposed and considered. Every drawing had a plunging hood and chisel-like prow, and an aircraft-type canopy with a tapering backlight. Front end proposals, however, had sugar-scoop headlight buckets or nonretractable Bug-eye Sprite-like pods on the hood. Rooflines all featured wide C-pillars, but with different contours. A one-piece wraparound hatch was most favored by designers.

Few sports cars, however, are "pure" designs. Reality, mostly in the shape of

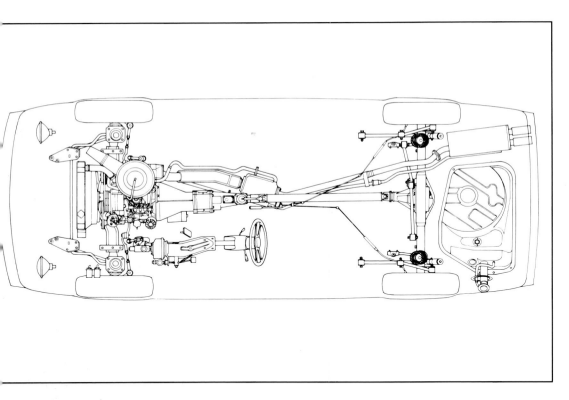

Although the RX-7's strut/live axle suspension resembled that of the RX-3, Toyo Kogyo engineers moved the engine 9.4in farther back in the chassis to achieve a 50/50 weight distribution. Mazda

The 12A engine as installed in the 1979 RX-7. The hose to the air cleaner lid is part of the vapor recovery system. Mazda

money but also in time, intercedes. Even the Porsche 911 makes concessions to its VW Beetle heritage. Project X605 didn't get the one-piece hatch but, in the interest of weight and cost, it was fitted with a frameless liftgate center section with two glass corner wedges. This backlight construction, Toyo Kogyo pointed out to reporters, offered more protection in a rear-end crash, had less distortion than a single curved piece of glass, and was easier to rainproof, though these were probably only collateral advantages to the cost factor.

The designers held their ground on lighting, though, and Project X605 was fitted with retractable headlamps. Extensive wind-tunnel testing made the body more than just good-looking: The final drag coefficient was cited as 0.36. Though not exceptionally slippery by today's standards, it was equal to that of the contemporary Porsche 924. Raising the headlamps increased the Project X605's Cd to 0.38, the same as Datsun's 280Z.

Compromises included suspension and steering that borrowed heavily from earlier Mazda sedans: MacPherson struts at the front, a live axle at the rear, and recirculating ball steering. The lower lateral arms of the front suspension were stamped and were located by trailing links. The coil springs were concentric, though tapered for tire clearance and to keep the tire center and kingpin axis close for reduced wheel vibration and more stable braking. Unlike many MacPherson systems, camber adjustment was provided, though it was through adjustments at the lower arm.

The rear suspension was similar to that of the RX-2, a live axle on coil springs with four trailing arms. But whereas the RX-2 had a Panhard rod, Uchiyama's chassis crew devised a Watt linkage, more expensive but providing better lateral location of the rear axle and, by nestling ahead of the differential, allowing a rear overhang some three inches shorter. The Watt link, interestingly, was adapted from competition RX-3s campaigned by amateur racers in Japan. Apparently racing *does* improve the breed.

The suspension offered more than adequate wheel travel for a sports car, with 7.3in at the front and 6.6in at the rear (with 2.8in in jounce in each case). The front struts contained conventional hydraulic shock absorber workings, but Kayaba gas-filled shocks using the De Carbon principle were fitted at the rear.

Hardly cutting edge, the recirculating ball steering box was another compromise, lifted from the RX-3. Even then *real* sports cars had rack and pinion. However, Mazda engineers found that some steering system play came not from the steering box but from steering shaft flex. They therefore increased shaft diameter from the RX-3's 25.5mm to 33mm and relocated the steering shaft bearings for additional rigidity. So while the RX-3 steering system would distort 19–33deg from a 7.2lb-ft (1kg-m) torque loading, Project X605's steering would deflect only 15deg. This was comparable to typical rack and pinion systems, which twisted 7 to 16deg under the same load. Power steering was not offered—it was considered out of place on a sports car. The steering system had a variable ratio of 17:1 to 20:1 and required 3.7 turns lock to lock.

Further compromise involved the brakes. While four wheel discs might have been expected, rear drum brakes were specified for Project X605, with discs at the front only. Other concurrent sports cars used this arrangement as well, however, including the Porsche 924 and

Ghost view shows how the rotary engine allowed the compact dimensions of the first RX-7.

the Triumph TR-7. The vented discs measured 8.9in in diameter (actually slightly smaller than the RX-3's), and the rear drums were 7.9 by 1.3in (slightly wider than the RX-3's) and were finned for cooling. Vacuum assist was standard, and the braking system split front and rear.

Although wheel diameters were only beginning to grow, wheel and tire combinations were a compromise as well. Mazda engineers specified styled steel wheels 13in in diameter and 5in wide. Base models wore 165HR-13 steel-belted radials, and a "plus one" tire sized 185/70HR-13, went on premium models. Optional were 13x5.5in aluminum alloy wheels.

Although the Mazda 12A rotary was specified for Project X605, Yamamoto and his engine engineers had not left it untouched. The little rotary was improved over the 12A used in the RX-3, and produced five more horsepower and three more pounds per foot of torque. The gain was mostly from detail changes, as the side-port design and two-stage, four-barrel carburetor would be retained, along with the thermal reactor for the U.S. versions. (Fuel injection was considered, but was not compatible with the thermal reactor, meaning that it didn't produce the over-rich conditions needed for Mazda's exhaust oven to work.)

The X605's 12A received a new apex seal design, "crowned" for more constant contact as the rotor waggled its way around the chamber. The seal's makeup was changed as well. The cast-iron was crystallized at the rubbing tip, by using an electron beam to "chill harden" the seal's edge.

The combustion chamber shape—the recess in the rotor face—was changed from a deep symmetrical configuration to a "leading deep recess." The leading spark plug (with three instead of two electrodes) was advanced 5mm, and the ignition alternated between single- and dual-plug operation. The trailing plug switched off at part throttle to assure a rich mixture in the exhaust to keep the thermal reactor cooking and emissions low. Both plugs fired at full throttle for power and at idle for smoother running.

Another emissions trick was a preheater for air injected into the thermal reactor. A heat exchanger in the exhaust stream warmed air from the air pump by jacketing it around an exhaust plenum. From there, the prewarmed air flowed back up along a double-tubed exhaust system, becoming even hotter before being shunted into the reactor itself. The heated air promoted combustion better than relatively cool air direct from the air pump.

Three transmissions were offered, two manual and one automatic. The manuals were a four- and a five-speed, the latter to be standard equipment on a premium version. Gear ratios on both transmissions were the same for first through fourth (3.674, 2.217, 1.432, and 1.000:1), while the five-speed had a 0.825 overdrive on top. The automatic was a JATCO three-speed. Final drive, regardless of transmission, was 3.909:1.

Although a number of interior styling configurations were tested, the winning dash design featured an instrument pod over the steering wheel with a prominent center console containing comfort and convenience controls, including ventilation and radio. The main instrument pod was dominated by a centrally placed 8000rpm tachometer. A 130mph speedometer was set to the right with a combination clock-fuel-temperature gauge on the left. The tach functioned as a voltmeter with the ignition on and the engine not running, for checking battery condition before starting the car. All gauges had white lettering on black, and red needles. A pair of bucket seats with bright plaid inserts was standard equipment.

The first prototype was completed in early 1977, with the first production models rolling off the assembly lines of the Ujina plant in March 1978. The model's United States debut was at the Mark Hopkins Hotel in San Francisco on April 24, 1978. A long-lead press preview

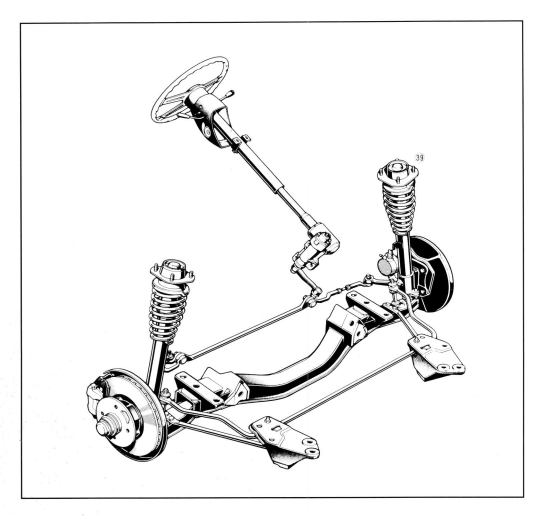

MacPherson struts formed the front suspension; the recirculating ball steering was an improved version of the RX-3 unit. Mazda

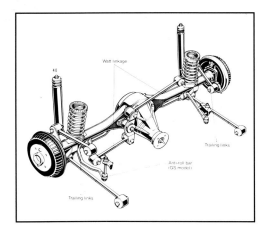

Mazda chose the more expensive Watt linkage over a simpler Panhard rod for improved axle location during cornering. Mazda

had been held earlier in Hiroshima, with the information "embargoed" until May issues for monthly publications.

For a country that had grown wary of the rotary engine, the RX-7—as Project X605 was to be publicly known—was a mammoth hit. Press response was uniformly positive, with niggling complaints made mostly to prove the maker's objectivity. But mostly it was, as *Road & Track* would call it, an enthusiast's dream come true.

No doubt, price was part of it. Base price was an amazing $6,395. More than a Fiat X1/9, sure, but that was an older design with a 0-60mph time of 16.3sec. The RX-7, on the other hand, clipped that sprint in 8.6sec, faster than the Datsun 280Z or the Porsche 924, priced $7,688 and $10,995, respectively.

The price of the base, or "S-Model," RX-7 was all the more impressive because it included AM/FM stereo radio with electric antenna, tinted glass, quartz clock, rear defogger, and Bridgestone 165HR-13 steel-belted radials. The GS sold for only $600 more and added a five-speed gearbox, 185/70R-13 steel-belted Bridgestone radials, a rear anti-roll bar, and a four-spoke steering wheel in place of the S-model's two-spoke wheel. Also included was an electric hatch release, a day-night inside rearview mirror, and miscellaneous trim bits, including the side molding not included on the base package (though often installed).

Options included cast-aluminum alloy wheels of a complex design made up of radial and circumferential ribs. Alloys were a common addition, listing for $250 and usually specified by dealers wanting to turn a little extra profit. They were worthwhile, however, if only for the extra half-inch of rim width. The intricacy of design, however, would add hours of cleaning time over the lifetime of any RX-7 so equipped.

Air conditioning was another option, priced at $525. The three-speed automatic added $355 but reduced the RX-7's acceleration—and the driver's heartbeat. California buyers spent $75 for that state's emissions package. A small ducktail spoiler was another option, recommended to anyone who regularly drove at 100mph—or who wanted to look like they did. A pop-up/lift-off sunroof added another $275.

The RX-7 was a bonanza for Mazda dealers, only recently restored to solvency by the subcompact GLC. Bullhorns may not have been necessary for crowd control at dealerships, but it definitely was a seller's market. Waiting lists formed, with prospective buyers having to wait months for delivery and paying as much as $2,500 over list.

Mazda had planned to produce 5,000 RX-7s per month, with 1,500 to 2,000 per month coming to the U.S. after an initial supply of 3,000. The relatively low number came from Toyo Kogyo's desire for quality over quantity and wish to "strengthen its dealer network, parts supply, and service." The ground swell of demand in the U.S. soon had the majority of RX-7s built being shipped across the Pacific. By the end of calendar 1979, Mazda had produced over 140,000 of its new sports car for export and domestic consumption. In the U.S. alone, 19,299 were sold in calendar year 1978, and 54,853 in calendar year 1979.

In 1975, Mazda's survival seemed

Would the RX-7 race? Of course, and Toyo Kogyo even displayed its own competition version at the RX-7's press introduction at Hiroshima. Road & Track

bound to anything but the rotary. That Yamamoto continued development of the engine was critical to the success of the RX-7, as was the belief of Akio Uchiyama and his fellow engineers and designers in the concept of a rotary-engined sports car. But who would have thought that a banker-turned-board member would have provided the organizational critical mass necessary to launch Project X605 in the first place?

Chapter 5

Evolution
The 1979 and 1980 RX-7

Racing Beat-prepared and Don Sherman-driven, this RX-7 went 160.393mph on the salt at Bonneville. Racing Beat

Road & Track had a reputation for sober but thorough evaluations, so praise such as that above was news indeed. Engineering Editor John Dinkel wrote, "This starts with the rotary's usual hushed idle that causes you to glance at the tach to make sure the engine is running. It continues with typically quiet and smooth and torquey rotary performance that has the tach twisting to the 7000rpm redline before you realize it. Then an over-rev buzzer jars you into reaching for the next highest gear."

Shifts were "light, quick, and a little notchy." The steering wasn't rack and pinion but provided "acceptable road feel and response." The RX-7 tended to toss its rear end in lane changes and lift and spin its right rear wheel on the skid pad. Low-speed corners provoked understeer. The rear felt light at top speed—the RX-7 could top its redline in fifth—so the optional dealer-installed ducktail spoiler was recommended. Dinkel noted

> *"To love the RX-7 is to drive it."*
> —Road & Track

that the RX-7 felt tight and rattle-free on rough roads compared to the Porsche 924.

Said Dinkel, "There's little I can think of that should be added to the car," calling the RX-7 "a car that will have as much impact on the sports car market and will acquire a cult of followers like no other sports car since the 240Z."

Car & Driver was no less enthusiastic and certainly not at a loss for words: "The two-place rotary rocket has more sex appeal than 'Charlie's Angels,'" said Don Sherman in a now-dated compliment, and continued "you'd gladly trade your favorite fantasy for an RX-7 after one quick test drive. A rotary engine humming eagerly through the gears is absolutely charismatic, and, if this doesn't ring your chimes, you'll be pleased to find the rest of the car in harmony with the melody under the hood. A low-slung RX-3 it's not."

Despite kudos thicker than all the stucco in Southern California, the RX-7 wasn't perfect. Sherman also noted that at high speed the RX-7's "steering lacked the surgical precision of the 924," although commented that the rear spoiler should help that. "On the high-speed

The 1980 RX-7 GS was elegant in black.

loop, we found throttle-induced oversteer easy to provoke, but the RX-7 is not a car you can pitch sideways and glide around corners like a figure skater." He advised that, "In its present trim, the RX-7 is best handled tidily and with respect for traction limits. The combination of 0.76-g roadholding (left and right average), and action-packed acceleration is more than enough to keep the competition at bay in any case."

Despite Sherman calling the RX-7 "a poor man's C-111," Toyo Kogyo almost immediately made changes. The prop rod for the hood on early RX-7s is on the right side, changed to the left side for later '79s. The lack of a passenger-side outside rearview mirror on early cars was corrected as well. The 1980 model year heralded a number of changes, but you could tell the difference only by looking at the rear bumper. The 1980 RX-7 had two rubber nubs added below the rub strip. The really sharp-eyed, though, might notice the radio antenna, eight inches longer for improved radio reception.

Inside the sure indication (for U.S. models) was the "safety" speedometer: For the 1980 model year, Joan Claybrook, National Highway Traffic Safety Administratrix for Jimmy Carter, mandated the national speed limit of 55mph be highlighted on speedometers calibrated to no higher than 85mph. Other evidence of a 1980 model is vinyl upholstery, optional for eating drippy ice cream cones.

Under the hood, electronic ignition finally reached the rotary engine. The old contact point distributor was replaced by a distributor with electronic timing, retrofitable to earlier models. The 1980 electronic ignition, however, was unique in that the igniters were separate from the distributor. Subsequent years would incorporate the igniters into the distributor body.

Two specials were released in 1980. The first was the Anniversary RX-7, in deep burgundy Mazda Renaissance Red paint with pinstriping and a medallion with the Roman numeral X, for Mazda's ten years on the U.S. market. Mazda said burgundy celebrated Mazda's "rebirth." A grand total of 8,000 RX-7s, 626s, and GLCs in Anniversary livery (with no count by model) were shipped stateside.

The Anniversary RX-7 was based on the "GS sunroof" model and included special upholstery on new "competition-style" seats, a four-speaker AM/FM sound system, inside (though not electric) remote controls for the side rearview mirrors, and the optional alloy wheels.

The other 1980 special was the curiously named "Leather Sport" RX-7. Introduced late in the model year (May 28), only 2,500 were made. Like the Anniversary, the Leather Sport RX-7 had competition-style seats, but upholstered with leather, which also covered the gearshift knob and steering wheel rim. The Leather

The Racing Beat speed-record RX-7 was powered by a peripheral-port 13B racing engine.
Racing Beat

The only exterior clues distinguishing the 1980 RX-7 from the 1979 were the two rubber nubs below the rear bumper's rub strip. Note the "Rotary Engine" badge next to the left taillight (1980 shown). Only 1979 and 1980 models wore this emblem.

Sports were also based on the sunroof model and had the four-speaker AM/FM sound system and remote adjustable mirrors, but had raised-white-letter, 185/70HR-13 steel-belted radials on gold-colored aluminum-alloy wheels. Colors for the Leather Sport included Solar Gold, Brilliant Black, and Aurora White. All were fitted with an "LS" badge on the B-pillar.

Sales for calendar year 1980 totaled 43,731. The RX-7 was a bona fide hit.

Competition 1979

At the press introduction in Hiroshima, Toyo Kogyo displayed its version of a racing RX-7, complete with rear spoiler, fender flares, and multi-color graphics. Even more importantly, preliminary homologation papers had been filed with the SCCA; the 1979 RX-7 would be eligible in C-Production in 1978. The RX-7 was also classified in IMSA's Under-2.5-liter class, though racers weren't expected until 1979. Rotary-engined Mazda sedans had a long history of road racing success in the U.S. as well as in Japan and the rest of the world. Toyo Kogyo would capitalize on this positive aspect of the rotary's

The GS models were easily identified inside by the four-spoke steering wheel.

A special Leather Sport version was offered in 1980, complete with leather upholstery, gold-colored aluminum alloy wheels, sunroof, and an "LS" badge on the roof pillar. Mazda

The factory 13B-powered RX-7 not only won the GTO class at the 1979 Daytona 24-Hour IMSA opener, it also finished fourth overall. Mazda

reputation in selling its new sports car. The almost-May introductiowas too late for preparing an effective race program with the new car, though Al Cosentino deserves "first IMSA" honors, for running an RX-7 at the 1978 Mid-Ohio IMSA race.

Ironically, the RX-7's first major competition success was not in a road race but on that alien Utah saltscape called Bonneville. The racers were the same team that had set a 160.393mph G-Production record in a 13B-powered RX-3: Racing Beat's Jim Mederer and Ryusuke Oku with *Car and Driver* technical editor Don Sherman driving.

For the Autumn 1978 effort, the RX-7 borrowed from MMA had its engine replaced by a Racing Beat-prepared peripheral-port 13B. "Prepared" is understatement. Racing Beat added a huge twin-throat 58DSF Weber carburetor and eccentric shaft-mounted Hall-effect electronic ignition boosting the engine's redline to 10,000rpm—at which point the rotor bearings went from marginal to sayonara. But ten grand was enough for Sherman to run an average of 183.904mph, shattering the three-year-old 167.208mph E/Grand Touring class record.

The RX-7's Daytona debut was equally auspicious. In the 1979 IMSA 24-hour event, the factory-built racers took first (Katayama/Terada/Yorino) and second (Bohren/Downing/Mandeville) in the Under-2.5-liter GT class, setting a class race lap record and placing fifth and sixth overall.

After the rather conclusive win, IMSA reclassified the peripheral-port engines, changing the weight per cubic inch required for the overall car from 0.9lb/ci to 1.1lb/ci. Thus for Sebring, an RX-7 would have to weigh 2,521lb, compared to 2,063lb at Daytona. Or, it could run side ports like the RS cars, or race in the GTX class with the turbo Porsches and prototypes.

Testing at Sebring, drivers found the extra weight made the rear axle flex so much that the brake pads dragged on the discs. Consequently, they ran the cars in Daytona trim in the GTX class and finished thirteenth (behind twelve Porsches) and eighteenth. Bob Bergstrom, who used a heavy-duty Ford rear axle under his peripheral-port RX-7, completed only sixty-four laps.

IMSA then relented somewhat, re-

Teammates Walt Bohren (#7) and Jeff Kline (#77) battled all season long, but Bohren won the 1980 GTU championship. Racing Beat

ducing the burden to 1lb/ci for the next race, which took place at a rainy Road Atlanta, where Bob Bergstrom took third. Mazdas then finished second through sixteenth at the Riverside six-hour enduro, but at the next race at Laguna Seca, Bergstrom made racing history by being disqualified after finishing sixth, having tossed the extra ballast from his racer right in front of the race marshals. It was only then, however, that IMSA returned to the original weight of 0.9lb/ci for the RX-7. The weight change didn't produce a Mazda sweep, however. In fact, Don Devendorf won the 1979 GTU drivers and manufacturers' championship for Datsun, while Bergstrom and Mazda finished second in their respective categories.

Competition 1980

Nine Mazdas, including an RX-2, were in the 1980 IMSA 24-hour Daytona enduro, and the Mandeville/Downing/Frisselle RX-7 lead in the results by finishing second in GTU. Pierre Honegger's new Holman and Moody-built RX-7 came in third, and an RX-2, piloted by Alan Johnson, Robert Giesel, and Bruce Nesbit surprised just about everyone with fourth in class. Lyn St. James with Mark Welch and Tom Winters continued the RX-7 parade finishing fifth in an RX-7. Mandeville, Downing, and Frisselle won GTU at the Sebring 12-hour in the old factory car, followed by a chain of Mazda victories: Walt Bohren (in a new Racing Beat RX-7) at Atlanta, Jeff Kline at Brainerd, Bohren at the Daytona Paul Revere, Golden State (Sears Point), Portland, and Mosport. At Mosport, the RX-7 clinched its first manufacturer's GTU championship, but a bad crash (from apparent heat exhaustion) put Downing in the hospital with a concussion. Bohren later wrapped up the drivers' championship as well. It was the beginning of an RX-7 tradition.

Mazda had hoped to interest competitors in SCCA C-Production by giving cars to selected racers, as well as parts and some monetary support. In fact, it was somewhat counterproductive, as the chosen cars and drivers wound up scaring off potential competition. At the 1980 SCCA runoffs, while Showroom Stock B was an RX-3 playground, the only factory supported RX-7s that were ready for the show were those of Stu Fisher and Don Kearney, both dnf'ing in the race.

Rod Millen, winner of the 1979 NARRA rally championship, switched from Nissan to an RX-7 for the 1980 SCCA Pro Rally circuit. He managed five open-class victories, but the title went to John Buffum in a Triumph TR-8. Millen finished second. Certainly, though, the SCCA would see more of the RX-7.

Chapter 6

1981
RX-7, Refined and Defined

There was something missing from the 1981 RX-7, but it was something that no one would miss: Mazda had eliminated the thermal reactor. The one device that was to make the rotary engine a smog beater was replaced in the 1981 Mazda RX-7 by a pair of catalytic converters. Ironically the "cat" had allowed the dirty old piston engine to meet the 1975 EPA emissions limits. Had the rotary engine gone over to the other side?

In any case, the question was moot. It was fairly certain that the rotary, contrary to earlier expectations, would not be the wholesale replacement of the piston engine. The very survival of the rotary was in question, and anyway, there's little sentimentality in engineering, especially where function is involved. The 1981 RX-7 was fitted with a "reactive exhaust manifold," a cast-iron manifold behind which two catalytic converters were mounted in series.

> *It was fairly certain that the rotary, contrary to earlier expectations, would not be the wholesale replacement of the piston engine.*

A "split-air" system directed air to either the exhaust port or the second catalyst, depending on where it would most aid emissions reduction. At low engine speed and deceleration, normally dirty phases for hydrocarbon and carbon monoxide production, air was sent to the exhaust port to oxidize the HC and CO in both catalysts. At middle engine speeds, the air control valve rechanneled air through a nozzle located between the two converters. With the exhaust port air stopped, the front catalyst took care of the oxides of nitrogen (NO_x), while the rear catalyst eliminated HC and CO. The rotary is naturally low in NO_x, so no exhaust gas recirculation (EGR) was used.

Also, under deceleration the air/fuel mix was cut off from the rear rotor chamber via a "shutter valve" in the intake manifold, the carburetor (still a two-stage four-barrel) feeding only the front chamber. To prevent drag from excessive vacuum in the rear rotor chamber, a "coasting valve" was opened admitting fresh air into the chamber.

Four-electrode spark plugs with tips

The "baroque depression" was gone with the 1981 RX-7's restyle.

Though equipped with a catalytic converter and electronic ignition, the 1981 RX-7 still had a carburetor.

Profile shows new contour of front and rear bumpers and new front spoiler. Mazda

closer to the combustion chamber were new, and, since it was no longer necessary, the trailing-plug shutdown system of the 1979–80 model year was eliminated.

The catalytic converters, introduced in Japan in the fall of 1979 (one year earlier so that experience could be gathered closer to home), allowed significantly leaner carburetor settings, since the cats did not need nearly as rich an HC diet as the thermal reactor. The new arrangement was called the "lean burn system," and it raised the combined EPA city/highway mileage from 20 to 24mpg for the five-speed equipped RX-7, a 20 percent increase! Oddly enough, this improved fuel economy was accompanied by the installation of a bigger fuel tank, 16.6 instead of 14.5gal. The improved fuel mileage in concert with the larger gas tank increased the driving range some 40 percent.

No change in horsepower was reported, but elimination of the cast-iron thermal reactor and the heat exchanger reduced weight by 85lb, despite the addition of the two catalytic converters. It was enough to knock a tenth off Mazda's claimed 0–60mph acceleration time, pushing it down to 8.6sec.

The four-speed was dropped for 1981, with all stick-shift RX-7s getting the five-speed. All RX-7s would also get a rear anti-roll bar, previously limited to upgrade models, with the diameter reduced to tame oversteer.

To the untutored eye the RX-7 was still the same old RX-7. But a 1981 parked next to a 1980 will reveal numerous exterior changes that cleaned up the styling of the first two years and also improved aerodynamics. The original front bumper had been a square-edged beam set into a slot in the RX-7's nose. It wasn't bad, but the 1981 restyle was a marked improvement, integrating a more rounded bumper into the lines of the car. The front spoiler, previously an add-on strip beneath the nose, was incorporated into the shape of the 1981 front end. The turn signals, now recessed, were still mounted in the front bumper and still vulnerable to park-by-ear damage.

The restyle of the nose lowered the coefficient of drag from 0.36 to 0.34; with the optional rear spoiler, drag was further reduced to a 0.32 Cd. Front-end lift was diminished as well, the coefficient of lift dropping to 0.12, compared to 0.18 for the 1980 model.

From the side, the new cars had a wider rub strip, integrated into the wider rub strip in the front and rear bumpers. The rub strip also incorporated the side marker lights, previously mounted below the rub strip. The new wider strip would be fitted to all RX-7s.

New optional and easier to clean alloy wheels were another 1981 identifier. And from the rear, the 1981 was easily distinguishable by the relocated license plate, which, formerly in what Werner Buhrer called a "baroque depression" in *Road & Track*, was now below the rear bumper. The taillights were restyled, incorporating a black plastic band placed between them. A final detail was the replacement of the earlier chrome MAZDA logos with black logos.

Inside, the most obvious change was the instrument panel. The binnacle was unchanged, but the instruments were reconfigured, which Mazda said, was inspired by customer requests. The innovative tachometer-cum-voltmeter was gone, though the tach still had center stage. The fuel gauge was set in the bottom of the tachometer, and a new combination gauge to the left contained a coolant temperature gauge, voltmeter, and an oil pressure gauge, the latter not on earlier RX-7s. The clock was moved to the center console. The gauges also were converted to trendy "fire orange" markings.

The console had its features and switches rearranged; the gearshift moved rearward 30mm and was shortened 50mm for a more direct and short-throw feel. A new heftier shift knob was used and the shifter boot redesigned. The handbrake lever was also changed.

The 1981 RX-7 was restyled with a smoother front bumper, lower front valance panel, and easier-to-clean wheels.

The radio was "new and improved," with three FM pushbuttons and a power antenna controlled by the radio on-off switch and the ignition switch.

A one-piece headliner and door panels were new, the former with sun visor indents, the latter with map pockets. Shiny trim was reduced, and interior lighting was improved. New across-the-board standards were an electric remote fuel filler release, intermittent wipers, and a stoplamp warning lamp. All RX-7s also had rear storage boxes located behind the seats.

Mazda continued the base S model and the upgrade GS model, though it gave the latter new equipment, including a headlamp reminder chime, illuminated ignition key cylinder, map lamp, halogen headlamps, a cargo area lamp, and remote-control door mirrors on driver and passenger sides.

The Leather Sport special model of 1980 was reprised in 1981 as a new top-of-the-line model called the GSL—in which leather upholstery was an option.

The most significant functional features of the GSL were the new rear disc brakes and a standard limited slip differential. The tilt-up/removable sunroof, optional on the GS, was standard on the GSL, as were the aluminum wheels. Also standard on the GSL were cruise control, a rear window wiper and washer, cloth door trim, a four-speaker AM/FM radio, power windows, a quartz digital clock, and special bucket seats with adjustable head restraints. And the radial tires, the same size (185/70HR-13) as on the GS, came with raised white letters on the GSL.

The reaction of the press to the GSL

The tachometer cum voltmeter was changed for 1981 by giving the voltmeter a separate needle.

was equivocal. The limited slip differential and disc brakes were a good idea, but writers wondered why they came only on a luxury model. "If it were up to us," wrote Bob Hall in *AutoWeek*, "we'd make the four-wheel discs an option for *all* RX-7s since someone on a budget might want the performance pieces in the low-cost, lightweight RX-7. Call us cynics, but our feeling is that the folks who would be predisposed to the luxurious GSL don't buy the car for performance as much as image."

Hall noted that although the horsepower rating hadn't changed, the catalytic converter-equipped rotary actually produced about five more horsepower. That this resulted in slightly slower acceleration was blamed on the extra weight of the GSL features. Other complaints involved Mazda's selection of low rolling resistance tires, which Hall felt had little lateral resistance as well.

Road & Track complained of too much rear brake bias, which actually lengthened stopping distances, and caused the rear disc brakes to lock up too easily. The magazine also said the rear axle had too much anti-roll, resulting in too much oversteer in the slalom.

The GSL certainly cost more. The base price (mid-year) for a GSL was $11,395. Air conditioning would add $610 to that, and leather another $650. Add the automatic transmission and the price soared to over $13,000. Surely the RX-7 was beginning to climb from the "budget sports car" class. Demand, however, showed little sign of slackening. For calendar year 1981, the model's fourth on the market, sales in the U.S. slipped by only a couple hundred, the final figure reported as 43,418. The RX-7 looked like a sustainable success.

Competition

Kent Racing was the "factory" RX-7 racing team for IMSA GTU in 1981; owner Dave Kent bought back the RX-7 he had built for Bob Bergstrom in 1979 and "borrowed" the Brad Frisselle car from its new owner Jim Mullen until a new car was built. Mullen would co-drive with Walt Bohren in endurance events, while Bohren ran alone in the sprints. Mullen got the Frisselle car back after Atlanta when the new Kent car was ready for Bohren. Lee Mueller, with four SCCA national championships to his credit, would be primary driver in the

\multicolumn{5}{c}{Comparative EPA Figures 1980 and 1981 RX-7}				
Model	Trans	Mileage	1981 MPG	1980 MPG
RX-7 (49-state)	5-speed	City	21	17
		Highway	30	28
		Combined	24	20
RX-7 (49-state)	automatic	City	19	16
		Highway	24	24
		Combined	21	19
RX-7 (CA)	5-speed	City	20	16
		Highway	30	27
		Combined	24	20
RX-7 (CA)	automatic	City	19	16
		Highway	24	22
		Combined	21	18

Bergstrom car.

IMSA required carburetors on all RX-7s, and mufflers, too, to everyone's auditory relief (unmuffled rotaries are *very* loud). Most Mazda competitors running the peripheral-port 12A were producing an estimated 270 to 300hp, and all—at least any experiencing any success—were equipped with big oil coolers and radiators. In this configuration the RX-7 was required by IMSA to weigh 2250lb. Side port engines had a minimum weight 200lb lighter.

Mueller and Bohren dominated the season, racking up eleven wins in sixteen races, with Mueller edging his teammate for the driver's championship. Mazda easily won the GTU manufacturer's title.

The Kent cars were not the only RX-7s in IMSA GTU competition. Jim Downing, with years of success in IMSA RS racing fielded a team backed by BFGoodrich to run their street tires in a race-tire class.

In truth, it took several months before the RX-7s were faster on street radials than the RS RX-3s, long familiar to Downing. The rear suspension was a modified Mazda competition live axle, while the front was mostly modified Datsun 280Z parts. There would be no wins for the street-tired RX-7s, but Jim Downing and Irv and Scott Hoerr placed third at Sebring, and Roger Mandeville and Amos Johnson claimed a second place finish at Riverside; impressive high-water marks against racers with race tires. It was only a beginning, however, for what would be the most successful car ever in IMSA competition.

The RX-7 began to see more success in SCCA competition, with John Hogdal's second in C-Production at the runoffs at Road Atlanta as the apogee. Meanwhile, Rod Millen took his RX-7 back to the woods to overwhelm the field and capture the SCCA national rally championship.

BFGoodrich sponsored Jim Downing to run an RX-7 GTU team on its street radials. Although the car, seen here competing at Mid Ohio, was less than successful in 1981, it would go on to become the winningest RX-7 in IMSA history. BFGoodrich

Lee Mueller won the 1981 GTU driver's championship in a Kent Racing RX-7. Mazda

Chapter 7

1982 and 1983
Detail Improvements

There was little for Mazda to talk about in 1982 regarding the RX-7. That's not surprising. The model was new for 1979, had a number of refinements for 1980, and was significantly revised for the 1981 model year. For 1982, all Mazda Motor of America claimed as new about the RX-7 was its tires: A change not in size but "model," from RD 106 to RD 207. Encouraging news, however, came from Japan: a turbocharged rotary debuted in Mazda's domestically marketed Luce and Cosmo series, the successor to the Cosmo RX-5. But in the U.S. the only new Mazda model was the diesel pickup truck.

Press reviews of the RX-7 were still favorable, but emphasis shifted to the RX-7's relatively low cost. *AutoWeek* ran a story entitled "Rehashing the Rotary: Mazda's Aging RX-7 Is Still a Great Value," and *Motor Trend,* in "Showdown of the Sudden Samurai: Supra vs. RX-7 vs. 280ZX" put the RX-7 in third place among this trio, saying that the "RX-7,

This 1983 RX-7 GS is unusual for having standard steel rather than the optional alloy wheels.

> "The RX-7 needs no apologies. It's a joy to drive."
> —Motor Trend *on the 1983 RX-7*

for its part, won grudging approval as the value winner, the most for the least money." The article went on to state that the RX-7 "delivers more horsepower per cubic inch than its adversaries, but there's not enough of it." The RX-7 clocked 1.5sec slower than the Supra in 0–60mph time, and almost as much slower when compared to the 280ZX. *Motor Trend* prognosticated a revision, however, for 1984: the return of the 13B engine with a positive displacement supercharger. It would prove half right.

Suspension-wise the RX-7 proved the adage that a good solid rear axle is better than a poor independent one, said *Motor Trend*. The magazine praised the four-link rear axle and its "fairly standard" Watts link. On the other hand, the "rubber bushings selected for the different linkages are fairly soft, in the interest of damping road noise and harshness, but they are a little too compliant for maximum axle location, evidenced in vague transients as weight transfer builds up to its steady-state value." Whew.

While the Supra actually clocked the fastest time around a road course, "the handling portions of this comparison set the RX-7 apart from the other two cars.

The 1982 Mazda RX-7 GSL; fender lip moldings were installed by dealers and owners.

It's character is much more pure sports car, which comes across as a somewhat harsher ride and a distinctly more nimble feeling . . . Even though it was the only car in the group lacking rack-and-pinion steering, the RX-7's recirculating-ball system produced excellent road feel, light effort, and crisp transient response." Despite the numbers, *Motor Trend* writers found the RX-7 was "the more gratifying car to drive."

The sales performance was at least as gratifying to Mazda; RX-7 sales actually increased by more than 12 percent. The final figure for calendar year 1982, a recession year no less, was 48,889 RX-7s sold in the U.S.

Competition, 1982

The RX-7 was certainly gratifying to drive in competition in 1982. Rod Millen found his hands full against the new four-wheel-drive Audi Quattro of John Buffum, but still managed four victories in SCCA national rallies, which was sufficient for a respectable second place in the standings. The good news was that a second place in rallying was the worst news RX-7 fans would get. In professional and amateur road racing, the news was very good.

The year began with a special factory GTO team at Daytona for the 24-hour race. Their RX-7 had a peripheral-port 13B engine running fuel injection for qualifying, but switching to carburetion for reliability in the race. Starting from fourth on the grid, the team of Katayama, Yorino, and Terada lead GTO until an axle bracket broke in the early morning hours. After a frantic forty-eight minute repair, the trio returned to the track, pulling out all stops. With two hours to go, they regained the class lead and never again to lose it. The class win, fourth overall, was a one-night-stand, however, as the car was sent back to Japan "for development."

In GTU RX-7s were equally impressive, with the two Kent team cars finishing sixth and seventh overall. The team of Bohren, Mullen, and *Motor Trend* writer Ron Grable took the GTU pole, but it was Mueller, Kathy Rude, and Australian Mazda racer Allan Moffat who took the GTU laurels. It was a history-making race: the third straight year that an RX-7 won GTU at Daytona, but also the first woman to win an IMSA race, and the first woman to win a major professional road race in the U.S.

It was Rude's last race in a Mazda, however, as Kent switched to Toyota for later 1982 races. BFGoodrich, meanwhile, shifted support to Porsche, leaving Downing and Mandeville with the ex-BFG cars as a natural match for Mazda to support. Converted to run on race tires, the ex-BFG duo finished third and fourth in class, driven by Mandeville, Amos Johnson, and Kline, and Downing, John Maffucci, and Mazda RS veteran Tom Waugh. The two trios finished in the same order at the Sebring 12-hour, but this time in first and second places in class.

The 1982 season was a repeat of 1981, though with Downing and Mandeville scrapping for the championship. Before Pocono, next-to-last of the nineteen races that season, Mandeville had a ten-point lead. But Downing and Maffucci finished first, while Mandeville and Johnson placed fourth after mechanical difficulties. Thus the points were tied for the Daytona finale. Downing won the Florida race and the 1982 season's driver championship when Mandeville's RX-7 blew an oil filter seal. Joe Varde, driving the former Frisselle car, finished third in the standings. Mazda, naturally, clinched the manufacturer's championship.

At the SCCA runoffs it was a pair of RX-7s on the front row of the C-Production grid. John Finger led the first seven laps and set the fastest lap, but Bob Reed won the race, giving Mazda its first win in a class often dominated by Datsun (Reed beat the 280ZX of Morris Clement by a mere 0.879 seconds). Finger finished third. Indeed it was a gratifying year for road racing RX-7s.

1983

Hallelujah, the 130mph speedometer was back! The Reagan administration's concession to reality eliminated Claybrook's folly, though a small dot at the 55mph marker highlighted the double nickel for the bureaucratic busybodies. Otherwise, changes were few for the RX-7 for 1983, with mostly upgrades and special models.

To hype the mostly unchanged RX-7, Mazda released the Limited Edition in March, and though this was mostly a trim package based on the GS model, it gave 5,000 buyers a Mazda out of the ordinary. The Limited Edition introduced a new color, Chateau Silver, accented by red pinstriping. Specially styled 5.5in-wide aluminum alloy wheels were mounted with Bridgestone 195/60HR15 radials, the biggest tire that Mazda had yet put on a production RX-7. Special badging on the rear roof pillar identified the Limited Edition, which included GS equipment plus

The fully optioned GSL had leather interior and an automatic transmission.

the pop-out sunroof, cruise control, "plush" red velour interior, air conditioning, and a special sound system with a four-speaker ETR/AM/FM stereo/cassette, Dolby noise reduction, and a seven-band graphic equalizer.

The changes to all 1983 RX-7s included new heater-ventilation controls, plus seats with lumbar adjustments and modified cushions. The pop-out sunroof be-

Jim Downing captured the GTU driver's championship in 1982 in the former BFGoodrich car. Mazda

Mazda sold 5,000 Limited Edition RX-7s in 1983, featuring Chateau Silver paint with red pinstriping, 195/60HR14 Bridgestone radials, and special 5-1/2in-wide forged-aluminum wheels. Mazda

came standard on the GS. Sound systems on the premium models were upgraded, with the GS getting a standard AM/FM stereo/cassette with six speakers and a manual tuner. The GSL added a graphic equalizer, electronic tuning, and a stick balance control.

Reprising its earlier "Samurai" story, *Motor Trend* again pitted the RX-7 against the 280ZX Turbo and Toyota Supra, adding the Mitsubishi Starion as a new and formidable competitor. Again, the RX-7, with its relatively meager 100hp, came in last in the four-way drag race, despite being the lightest. Its top speed of 117.5mph barely edged the Supra's 117.0mph, though the testers suspected that this Supra might have been slightly off song. The RX-7 was most predictable, though limited by its relatively narrow tires, on the skid pad. The slalom was a tie between the biggest tires (Supra), shortest wheelbase (280ZX), and most stable chassis (RX-7), while the Starion displayed the best combination of all three. The Mazda, despite its rear discs, wasn't the best in braking, as the rear wheels locked.

Despite not topping any of the test categories, *Motor Trend* said, "The RX-7 needs no apologies. It's a joy to drive. That unique engine creates some tingling vibration, particularly when working hard, but it loves to rev, like few motors available today. Peak power comes up at 6000rpm, the redline at 7000, and it makes no complaint winding past 8000. Power doesn't even fall off very much. With the smooth 5-speed box, this is among the most fun-to-use powertrains you can buy today." The magazine still picked the Supra for its choice among the four, with a nod to the 280ZX Turbo's power and Starion's racetrack performance, while acknowledging that the RX-7 "on a dollar-for-dollar basis . . . obviously makes a strong case for itself." Perhaps if the RX-7 had more power?

In Japan it did. The 12A turbo engine was made available in the RX-7 for 1983. The domestic naturally aspirated RX-7 was rated at 130hp JIS (Japan Industrial Standard, but the Hitachi HT 18S-BM turbocharger, using exhaust pulses plus conventional exhaust pressure, bumped output to 165hp JIS for the RX-7 Turbo GT-X. *Car and Driver*'s report included no acceleration times but noted that "a blower has little effect on the classic low-speed torque deficit of the RX-7, but a second stage seems to light off once 2000rpm is reached. The rotary rocket then shoots impressively to its 7500rpm limit." A top speed of 140mph was observed on the car's speedometer (at 6800rpm in fifth gear).

The GT-X drivetrain was strengthened and the suspension revised. Rear roll stiffness was reduced by lowering the mounting points for the lower trailing links on the rear suspension. Limited slip was standard as were 205/60VR-14 radials fitted to larger diameter wheels. All four corners had eight-position manually adjustable dampers. Bigger brakes along with power-assisted steering with a quick 15.8:1 ratio were also included. Unfortu-

Roger Mandeville took the 1983 IMSA GTU driver's crown. Mazda

nately, it was only for statesiders to read about, as there were no plans to export the GT-X to the U.S.

RX-7 sales rose again, nearing the 1979 rate. The U.S. sales for calendar year 1983 totaled 52,514, countering the sales taper normally expected of a five-year-old car.

Competition, 1983

There was certainly a lot to read about when it came to racing RX-7s: IMSA in both GTO and GTU classes, more SCCA competition, and a unique RX-7 for rallying.

The rally RX-7 was Rod Millen's. Frustrated by the traction advantages of John Buffum's Audi Quattro, Millen used the off-season to build a four-wheel-drive RX-7, mostly as an experiment. The front-wheel-drive suspension and drivetrain from the Mazda 626 were adapted to the tube-frame RX-7, a power take-off unit from Weismann's Transmission Systems Research and Development fitted to the rear of the transmission.

The conversion to four-wheel drive raised the RX-7 a mere three-quarters of an inch; Millen reporting no deterioration of handling. The rotary made about 300hp, about 50hp less than its German-built rival, but also weighed less. The conversion also added only about 100lb. Braking and overall stability were

The interior of the 1983 Limited Edition featured red velour seats. Note the speedometer calibrated to 130mph. Mazda

Although changes were few for 1983's RX-7, a welcome one was the return of the 130mph

Pacific Avatar and the Topless Californians

Just looking at the first-generation RX-7 was enough to provoke speculation about Mazda's rotary sportster's appearance as a convertible. The cockpit styling of the first-generation RX-7's roofline had a break between the body and the roof, leaving a body shape that, at first consideration, seemed a natural for a convertible conversion. And weren't "real sports cars" convertibles anyway?

With Kalifornia Konvertibles being made of everything from Mustangs to Countaches, the thought of the RX-7 as a convertible, without doubt, occurred to many. Although he may not have been the only one to take the top off an RX-7, Al Dooley possibly did it the most and maybe the best.

Dooley, sales manager for a Mazda dealership in 1978 when the RX-7 debuted, found the car's clean lines irresistible. He decided he would make convertible RX-7s if Mazda wouldn't.

Setting up shop as Pacific Avatar in Garden Grove, California, Dooley started each conversion by lopping off the top. The fastback RX-7, however, presented more problems than a notchback would have. The RX-7 had no rear deck, so Pacific Avatar constructed one from sheet steel, and continued the contour of the RX-7 coupe body. Making the entire rear deck wasn't necessary, though, as Dooley serendipitously discovered that the Mazda 626 rear deck had the same curve as the RX-7's. Dooley needed merely to adapt the 626 trunk lid, complete with its factory-stamped reinforcement on the underside. The biggest problem was that the lip on the lid was too short, but this was corrected by adding a strip of steel and finishing it, thus giving it the appearance of a factory job.

Razing the roof removed a lot of structural rigidity. Pacific Avatar strengthened its topless Mazda by adding an X-shaped steel brace under the body, quarter-inch steel plates on the rocker panels, additional longitudinal bracing behind the seats, and a loop around the back of the body opening. Later cars had a plastic material added inside body panels for additional sound deadening.

The top, which had a striking resemblance to that of the Mercedes-Benz 450SL, folded into a well behind the seats and under a metal cover, leaving some storage room behind the seats even with the top down.

Although he made some "vanilla" cars, as Dooley calls simple convertible conversions, more had additional modifications, including "wide-body" fender flares with wheels and tires to match, and brake upgrades and suspension modifications. Most popular were turbo kits. Prices started at

The convertible conversion by Pacific Avatar was a natural fit with the RX-7's clean body shape. Al Dooley

Use of the Mazda 626 trunk lid produced a professional final product. Only a slight extension on the bottom lip was needed. Al Dooley

$25,000 for a complete "base" convertible, with the sky the limit for a full custom convertible.

Florida was the biggest market for Pacific Avatar's convertible RX-7s, although Dooley says the cars were also popular with those in the California movie industry. One even had a supporting role in the low-budget film, Smokey Bites the Dust.

According to Dooley's records, Pacific Avatar performed 126 conversions between 1979 and the last convertible car bound for Maryland in 1987. The cars were sequentially numbered and identified by a data plate in the engine compartment. When the second-generation RX-7 was introduced, Dooley investigated a convertible conversion, but when he received word that Mazda had its own in-house version planned, Dooley dropped plans for a new convertible of his own, building out a few more first-generation cars as demand decreased and ended.

The Mercedes SL-type top eliminated the driver's blind spots. Al Dooley

actually improved, and the RX-7 also had a higher top speed than the Audi.

"We did some checking on Buffum's Quattro," Millen explained before the start of the 1983 season. "We found out that it was five seconds faster into the first corner [from a "dead stop" starting point] on every stage." Millen's overall stage times in the two-wheel-drive RX-7 were usually very close to Buffum's, usually less than five seconds. And that was what Millen had hoped to make up with the four-wheel-drive system.

"We only get about six feet of 'scratch' off the line now, whereas before we'd spin the tires through the first four gears," said Millen. Gearing had to be changed as a result, as the four-wheel-drive RX-7 bogged as soon as it got traction. Another problem from the added traction was that the racing clutch, which had previously survived six rallies, barely lasted through a day of testing; a beefier twin-plate clutch was installed to handle the punishment. Peter Weismann also installed stronger racing gears in the transmission, just in case.

Millen's "experiment" worked so well that he decided to campaign it. He managed two wins, good for a first year's effort, but still only enough for a bridesmaid's finish in the championship behind Buffum.

An MMA-supported effort in IMSA GTO was also down on power against its rivals, though in this case 300hp RX-7s were up against 600hp Porsches. Working against these tremendous odds, Racing Beat made several modifications to their fuel-injected, peripheral-port 13B-powered racers. Racing Beat fine-tuned the body work by rigging an RX-7 with 31 manometers, thin U-shaped tubes filled approximately half full with red instrument oil and used for measuring vacuum or pressure. While the car was run at top speed, a remote-triggered Polaroid camera was used to take pictures of the manometers. Subtle adjustments of the bodywork were then made to decrease drag, all without a proper wind tunnel.

Another innovation was a multi-adjustable double A-arm front suspension that could be altered in minutes in response to changing track conditions.

And for Daytona, Racing Beat secured the services of Pete Halsmer, an Indy car driver who would become a Mazda fixture, and Rick Knoop. Bob Reed would be rewarded for his C-Production championship with a co-driver stint at Daytona.

The power deficit meant that enduros would have to be driven like sprints, equalizing the relaxed long-distance pace of the competition with the reliability of the Mazda rotary at full tilt. Well, Cinderella stories do happen. At the Daytona 24-hour race, the Racing Beat team started ninth (out of thirty-seven) in GTO and twenty-second overall. But by shortly after midnight, Halsmer, Knoop, and Reed were leading their class and were third overall. It was a position not surrendered even for a collision in the dark—requiring body repairs—nor for a rainstorm that halted the race for almost an hour, nor for conditions so rugged that only twenty-nine out of seventy-nine starters finished the race. It was the best-ever finish for Mazda or any Japanese manufacturer at the Florida classic.

At Sebring, the team, with Halsmer and Knoop driving, qualified sixteenth and was leading overall in the largest (eighty-four cars) and most diverse race ever for the airport venue, when a brake rotor came apart ten and a half hours into the 12-hour race. They had sufficient laps, however, to be credited with an eighth-place finish in GTO, and twenty-first overall.

The Racing Beat GTO racer would celebrate one more victory in 1983, at the Mosport 6-hour enduro. Qualifying fifth in GTO, Knoop and John Morton pushed the underpowered RX-7 to the head of the class during the second hour. At the

The Racing Beat-prepared Halsmer/Knoop/Reed RX-7 placed first in GTO at Daytona despite having half the horsepower of its rivals. Mazda

end of the race, the nearest competitor was six laps back.

But running as hard as they could in the enduros left nothing for sprints, and when their competitors opened up for the shorter races, the Racing Beat RX-7 simply didn't have a big enough motor for GTO.

Meanwhile GTU became the RX-7's backyard. In fact, the RX-7 had such an advantage that Dave Kent, with no factory backing for 1983, struck a deal with Firestone to run street tires—not high performance radials but $36 passenger tires rated for 110mph. Lee Mueller, Terry Visger, and Hugh McDonald were the brave souls piloting Bob Bergstrom's old racer, not only finishing at Daytona but winning the class. Four out of the next five GTU finishers were Mazdas as well.

At Sebring, the Mike Meyer Racing RX-7 posted its first GTU win with Jack Dunham, Jeff Kline, and Jon Compton driving, with RX-7s placing second and third. Roger Mandeville became the first winner ever on the Miami street circuit, the 1983 GTU race being the first ever held at that venue. Mandeville used a sideport 12A in the tube-chassis car, trading horsepower for lightness. The theory worked well enough for five race victories and the GTU championship. Jim Downing, busy building a Mazda GTP car, placed second in the year-end standings, and Mazda got its fourth consecutive GTU manufacturer's championship.

Rod Millen entered a four-wheel-drive RX-7 to counter Buffum's Audi Quattro. Mazda

Millen's 4WD RX-7 used Mazda 626 front-wheel-drive components. Mazda

Chapter 8

1984–1985
The GSL-SE and Other Changes

"We search constantly for ways to refine and fine-tune our products to keep them among the most fun to drive in the world. The GSL-SE has been developed as an outstanding performance package that still represents the kind of exceptional value for the dollar that has made the RX-7 the success it is. We believe, however, that improved performance is valuable only when it is in harmony with a vehicle's other components."

It sounds like *Zen and the Art of the Rotary-Engined Sports Car*. It was, however, Yoshiki Yamasaki, president of Toyo Kogyo, introducing the new 1984 Mazda RX-7 GSL-SE. While *Motor Trend* had predicted a supercharged rotary—something Yamamoto's engineers had been developing—and others expected the turbo already on the domestic market to be loaded on the boat to America, Toyo Kogyo instead dusted off and spruced up the 13B rotary. And for mechanical harmony, the 13B RX-7 got the GT-X's chassis modifications.

> "For the money there still isn't a better fling-about, redline-hungry, tire-smoking sports car to be had."
> —Car and Driver *on the 1984 RX-7 GSL-SE*

It was called the GSL-SE, and if not *all* new, it was thoroughly revised. Most obvious was the engine: with exactly the same displacement as the earlier 13B, its measurements were identical, but little else was the same. Indeed, though the 13B's rotors were 10mm wider for 14.1 percent more displacement than the 12A, it developed 33.6 percent more horsepower and 24 percent more torque.

The secret was mostly in breathing and electronic fuel injection. While the staged primary and secondary ports of the 12A were retained, an additional port was added by enlarging and splitting the secondary port, making it a six-port engine. Mazda called it 6PI, for six-port induction.

The multiple ports were to maintain intake charge velocity at all throttle openings and engine rpm. The primary ports, located in the side plate between the rotors, and the secondary ports, on the end plates, were controlled by staged throttles. The new third ports in the end plates

The pop-out sunroof was standard on the GSL-SE.

Transmission/Final Drive Ratios					
		1983 12A	1984 12A	1984 13B	
Trans. ratios	(1st)	3.67:1	3.62:1	3.62:1	
	(2nd)	2.22:1	2.18:1	2.18:1	
	(3rd)	1.43:1	1.42:1	1.42:1	
	(4th)	1.00:1	1.00:1	1.00:1	
	(5th)	0.82:1	0.81:1	0.76:1	
Axle ratio		3.91:1	3.91:1	4.08:1	

were opened by rotary valves, but only at high rpm in response to exhaust pressure. The extra port area for high rpm allowed the main ports to be optimized for mid-range rpm. Electronic fuel injection (called EGI by Mazda, for "electronic gas injection"), with one injector in each primary port, permitted further fine tuning.

Yamamoto's engineers also maximized the ram tuning effect of the intake tract, an old piston-engine trick but with a new rotary-engine twist. A rotary's ports close faster than a piston engine's, and the intake pulses are much stronger, including the reverse pulse when the intake charge hits a closed port. Mazda harnessed the return pulse with a surge tank that directed it down the opposite intake tract, timed for ram effect, and into the other chamber. Mazda called it DEI, for Dynamic Effect Intake.

A secondary ram tuning came from what Mazda referred to as Residual Gas Effect: exhaust pressure left in the chamber when the intake ports opened would send a pressure wave up the intake tract. Carefully calibrated and timed, it boosted ram effect. The intake tract was lengthened for maximum effect at low rpm, where the rotary needed the most help. And anyway, at higher rpm, the engine

A new three-spoke steering wheel was the most obvious feature of the refurbished 1984 GSL-SE interior.

outruns the resonance effect and benefits more from larger port area. The bottom line was 135bhp at 6000rpm, with maximum torque of 133lb-ft at 2750rpm—down from 4000rpm for the 12A, for a much broader spread of power. For a smoother and more constant 800rpm idle, Mazda added a "bypass air control system," which compensated for power drains such as air conditioning and power steering.

The new 13B also gained injectors feeding oil directly onto the trochoidal surface of each combustion chamber. Lubrication was enhanced further by the micro-porous chrome plating of the trochoidal surfaces, which retained more lubricant and reduced wear.

The extra grunt of the new 13B warranted a stronger clutch, which was more heat resistant and clamped with ten percent more force, but clutch size wasn't changed. Gear ratios, however, were changed to match the new engine's power curve. The GSL-SE's five-speed manual (no optional automatic) was given a taller overdrive fifth-gear ratio that, even considering the lower rear axle ratio (4.07:1 for the GSL-SE, 3.91:1 for 12A-powered RX-7s) yielded lower engine speeds in fifth gear. On the other hand, for acceleration the lower axle ratio more than offset the marginally higher first- through third-gear ratios.

Chassis changes kept handling in harmony with performance. New twelve-spoke 14in wheels wore 205/VR14 P6 Pirellis, and, as on the GT-X, the mounting points for the lower rear trailing arms were lowered by 20mm (0.8in) to provide roll understeer to reduce the RX-7's tail-happy tendencies.

More power also meant more braking was needed, so Mazda fitted bigger stoppers, 9.8in vented discs up front and 10.1in solid discs at the rear (the front brakes had more swept area and stronger braking effect despite their smaller diameter). A larger master cylinder was fitted too, though neither caliper nor pad size was increased.

Power steering, a $300 option, had speed-dependent variable assist, which eased parking but maintained road feel at speed. The power steering also had a faster ratio, a straight 15.83:1 compared to the variable 17:1–20:1 ratio of the standard steering.

Interior changes for all RX-7s included a new instrument panel with same-size speedometer and tachometer, with a smaller combined gauge on either side for oil pressure, voltage, coolant temperature, and fuel level. Also new in the interior was a three-spoke steering wheel and a new shift knob. The console got a shiny silver insert, along with, for the GSL-SE, an oversized AM/FM/cassette stereo with a graphic equalizer and stick balance control. Rotary knob ventilation controls for the standard air conditioning were also new. The pop-out sunroof was standard on the GSL-SE, as were an eight-way adjustable driver's seat, power remote rearview mirrors, cruise control, and power windows. Leather was an option.

Only special wheels clued curbside spotters to a passing GSL-SE. The only badge was a GSL-SE emblem just above the left taillight.

Perhaps that's all the driver of a smaller-engined RX-7 would see of a moving GSL-SE, for which Mazda claimed a 0–60mph time of 8.0sec, and a quarter-mile time of 16.1sec. When *Motor Trend* ran yet another "Samurai" comparison test (against the Mitsubishi Starion ES, Nissan 300ZX Turbo, and Toyota Supra—only the last unchanged from 1983), it found the GSL-SE significantly improved against its Japanese rivals. Instead of dead last in acceleration, it was "nibbling at the heels of the almighty 200hp 300ZX."

"With the new injection of horsepower hormone," said the magazine, "the RX-7 came in second in all the acceleration tests, third in top speed, second in lap speed—a category we didn't have last year—and absolutely dominated the steady-state handling test with a blistering 0.88g around our skidpad vs. 0.77 for last year's edition."

Motor Trend claimed the only car better on the skid pad was the new Corvette, and that only by a whisker. The 13B engine bumped the RX-7's top speed to 125mph (confirmed by other

New badging on the rear deck identified the 1984 GSL.

tests and better than Mazda's claimed 120mph) and allowed it to edge out the Starion and Supra in acceleration. When braking from 100mph, the rear axle would hop, but the magazine conceded that most people wouldn't drive this way anyhow. The magazine didn't name a Samurai winner—the 300ZX was fastest around the race track—but called the RX-7 "the undisputed hell-raiser. It has this light-on-its-feet, agile feel to it that always seems itchy for a fight."

Car and Driver was smitten by the GSL-SE, too. "For the money," it concluded, "there *still* isn't a better fling-about, redline-hungry, tire-smoking sports car to be had." The new engine was heavily praised, both for its engineering elegance and the way it worked in the car. Drivers at the magazine found that when driven hard, the rear end could still be hung out, but it was much more controllable than before. In fact, the magazine criticized little more than the recirculating-ball steering, called vague on center, and the shiny panel on the console.

Road & Track compared eight competitors for the new SCCA Showroom Stock GT class, including the GSL-SE, in its August 1984 issue. Corvette proved class of the field, both in *Road & Track*'s test and SCCA competition. The RX-7 found itself in deep water in this match-up, which included not only the 'Vette but (again) the 300ZX Turbo, the Mustang GT, Camaro Z28, Porsche 944, Firebird Trans Am, and the Dodge Daytona Turbo Z. In laps times around Willow Springs International Raceway, that's how they finished, with the RX-7 just ahead of the Dodge. The magazine's drivers were disappointed with the GSL-SE's track performance, finding "judder and hop" under hard braking. Oversteer, particularly on trailing throttle, "limits the driver's options." Nor did relatively narrow rims and tires allow maximum advantage from the RX-7's weight advantage. The conclusion: "Fun . . . but not terribly fast." Ouch.

Despite the attention the GSL-SE was

The RX-7 GSL-SE, powered by the 13B rotary engine, debuted for the 1984 model year.

receiving, the "old" RX-7 received a number of improvements as well, including exactly one more horse for the U.S. legal version of the 12A. The power gain came from a more efficient monolith catalytic converter, so it was ironic that EPA combined fuel mileage dropped slightly, from 23 to 22mpg. A four-speed overdrive automatic with a lock-up torque converter improved fuel economy and performance for those preferring shiftless driving, optional with the GS or GSL versions. The variable-assist power steering,

This 1985 base model is equipped with optional alloy wheels.

standard on the GSL-SE, was an option on the GSL. All RX-7s also got the rear suspension modifications of the GSL-SE.

Other across-the-board changes included the new instrument panel and rotary dial ventilation controls, as well as improvements to the ventilation system, which included a four-speed fan. Wiper blades were longer, and the steering wheel and shift lever knob were new. And finally, locks and lights were added to the storage bins behind the seats.

The 1984 model year was the next to the last for the first-generation RX-7, and with an all-new sports car in the works, Mazda left the RX-7 virtually unchanged for the last roundup in 1985. About the only notable mechanical difference was that all wheels, steel or aluminum, had 5-1/2in-wide rims.

Sales remained strong, and, in fact, 1984 was a record year for RX-7 sales in the U.S., which at 55,696 surpassed even 1979. Mazda's U.S. car sales, however, after a record-setting year in 1983, slipped slightly. But Mazda gained one dubious distinction: having installed a 61hp diesel four in the 626, it was the only car manufacturer to ever offer diesel-, gas-, and rotary-engined automobiles!

In calendar year 1985, RX-7 sales edged down to 53, 810, due in part to the general knowledge that a new RX-7 would be coming in the 1986 model year. The new model's strong sales in the fall of 1985, however, somewhat countered these earlier lower sales figures.

In 1984, the Hiroshima-based company finally laid to rest the final vestige of its cork-producing origin. Long known for its production of Mazda automobiles, Toyo Kogyo Co. Ltd. changed its name to Mazda Motor Company. And in the same year, Kenichi Yamamoto, who had become known as "Mr. Mazda" for his work with the rotary engine, was named president of Mazda Motor Corporation.

The 1985 RX-7 GSL-SE was little changed from its predecessor.

Competition 1984

Perennial Mazda racer Jim Downing switched to an Argo JM-16, though Mazda-powered, to compete in IMSA's GTP class. Meanwhile in IMSA GTO, Roger Mandeville signed on as a David against the V-8 Goliaths. Mandeville's 1983 GTU championship RX-7 became his 1984 GTO challenger by the simple expedient of an engine swap, the installation of a peripheral-port 13B. Down on horsepower, Mandeville and regular co-driver Amos Johnson, and sometimes-co-driver Danny Smith, always found a way to finish: third at Daytona, fourth at Miami, second at Sebring and Road Atlanta. At the Riverside 6-hour enduro, Mandeville outlasted and outran everyone in GTO and also took the GTO points lead. Another win came at a rainy Lime Rock. The only nonpoint finish came after clinching the championship. With no other RX-7 in GTO, Mazda didn't get the manufacturer's title. But driver Mandeville's championship certainly gave bragging rights to Mazda, and for Mandeville it meant a driver's championship in three IMSA classes: RS, GTU, and GTO!

Meanwhile, in GTU, Mike Meyer Racing campaigned the old Racing Beat GTO car adapted to GTU as Mazda's factory team. Ira Young, owner of the Malibu Grand Prix miniature race tracks, bought the ex-Downing GTU RX-7 and hired Jack Baldwin as full-time driver; Young was the co-driver in the enduros. Both were new to GTU, though Young had raced an RX-3 in RS.

Not only did the Malibu Grand Prix team win their first time out, they did it at the grueling Daytona 24-hour enduro, Baldwin and Young joined by Bob Reed and Jim Cook. They followed with another win at Sebring, and though they didn't win all the races they entered that season, they never failed to finish, placing outside the top five only once.

Thirteen races into the seventeen-race schedule, Mazda had its fifth consecutive GTU championship. Jack Baldwin tied up the driver's title at Michigan, race number fifteen. Official factory drivers Jack Dunham and Jeff Kline finished second and third in the driver's standings.

Although the 1984 SCCA runoffs at Road Atlanta will be remembered mostly for the Great Crash that saw the destruction of eighteen GT-1 cars in a sudden rainstorm, GT-2 was decided by the empty gas tank in John Hogdal's RX-7 rather than Hogdal's last lap duel with 280ZX-mounted Morris Clement. In the final laps, the two swapped places while

Jack Baldwin dominated IMSA's GTU in the ex-Downing RX-7. Mazda

After leading the standings in mid-season, Rod Millen finished in the bridesmaid position in the 1985 SCCA/Bridgestone Pro Rally series. Mazda

staying ahead of defending champion Dr. Robert Reed—whose Mazda was miraculously put back together after a practice crash—and RX-7 hot shoe John Finger, who was hampered by a balky gearbox and a backup engine. The SCCA had awarded the 280ZX a 200lb minimum weight break after Reed hadwon in 1982 and 1983, making the Datsuns more competitive. Unfortunately, the battle was not won with firepower but logistics, and Hogdal sputtered while the Clement took the checker.

Alas, Rod Millen also was destined for a runner-up slot. He built a new four-wheel-drive RX-7 for 1984's national rally championship drawing on lessons learned from his first four-wheeler—including limited slip at both front and rear and 150lb less weight. An oil cooler inside the rear spoiler was a particularly interesting modification.

Millen and his RX-7 showed more consistency than arch-rival Buffum and his Audi. Though Millen had only three wins to Buffum's five, he led in points going into the final event. But with neither the driver's nor manufacturer's championship clinched, Audi defended its U.S. rally reputation by bringing in 1983 world rally champion Hannu Mikkola as a reinforcement. Millen countered by putting brother Steve in his 1983 as backup. Alas, Mikkola won and Buffum was runner-up, and although Brother Steve finished fourth, Rod had broken his transmission for his only dnf of the season. It was bridesmaid again for Rod Millen.

On the other hand, Richard Kelsey, an employee of Millen's Newport Rallying, entered a 12A-powered RX-7 in the SCCA's new Production GT rally class. Kelsey not only lead the standings all season but also helped Mazda notch up a manufacturer's championship in GT.

Competition 1985

No doubt, Mazda had been impressed by Jack Baldwin and the Malibu Grand Prix GTU team, and provided the team with some factory support. But it must have given the Malibu GP team some sort of overdog jinx: They dnf'ed at Daytona, the first race of the new season and their first failure to finish in the venerable RX-7 racer.

The Daytona underdogs had to be Amos Johnson's Team Highball, which didn't begin building its new tube-frame

GTU RX-7 until late December when it was also prepping IMSA RS cars for the new season. Johnson had never before built a rotary engine nor even been able to test the car before arriving in Daytona. But in the race, Johnson, who with Jack Dunham and Yojiro Terada started fifth, moved relentlessly into the lead, ending up forty-one laps ahead of the next GTU car.

Baldwin would not be denied, however, and returned to take Miami, next on the calendar. At Sebring, Baldwin, with Jeff Kline co-driving, would have a "not running at finish" finish, but wins at Road Atlanta and Riverside gave him the lead in the point standings. High point finishes combined with a victory at Watkins Glen clinched Baldwin's second consecutive driver's championship. It was the car's third GTU championship, and for Mazda, the GTU manufacturer's championship was its sixth in a row. Plus, Baldwin's win at Elkhart Lake made the RX-7 the winningest model in IMSA history, its sixty-seven first-place finishes topping the sixty-six credited to the Porsche Carrera. "There is no question that this was the single most important victory of the season for us," Baldwin claimed after the race.

Mike Meyer Racing put a young Scott Pruett in its GTU RX-7 along with Paul Lewis. Although former kart driver Pruett scored no wins, he was off to Ford at the season's end for a factory ride and the beginning of a successful racing career.

For IMSA GTO, Roger Mandeville built a new RX-7 in hopes of a second class championship, and turned his "old" car over to Danny Smith, making them a two-car team. Ford's better idea, however, was to back a Jack Roush effort in GTO, which, as a preview, won the first five events of the season. Danny Smith did snare a single victory at Charlotte when the Mustangs and the factory Toyota were all slowed by accidents. At season's end, Smith was fourth and Mandeville sixth in the GTO standings.

At the SCCA's runoffs at Atlanta, the RX-7 was the dominant force in GT-2 with six out of the top twelve places on the grid. Pole-sitter John Finger had used up his RX-7 in qualifying, however, so it fell to John Hogdal to bring home another national championship for the RX-7. Finger finished fifth, and Don Kearney brought his RX-7 in sixth.

The runoffs were only the second race for Hogdal's new Mike Lindorfer-chassised car, which had been the talk of the paddock. Hogdal later admitted to "sandbagging" during the qualifying rounds, shifting 600rpm early. If, said Hogdal, he had too easily outrun the field, the SCCA would have just added weight to all RX-7s for the following season.

After 1984's fifth-place finish, Dr. Bill Schmid set a new class record in qualifying his RX-7 on the GT-3 pole. But the race had just started when his motor went soft. Schmid held on for a fifth-place finish though, and Bill Van ended up sixth in the only other RX-7 in the race, while Bruce Short and Howard Coleman placed second and fourth in RX-3s.

In Showroom Stock A at the runoffs, the highest placed RX-7 finisher was David Sikes in eighth place. The RX-7s were out-powered in the new-for-1985 SCCA/Playboy United States Endurance Cup series, though the Falcon/Grissom team of Tom Mankin, Scott Grissom, and Tom Stuart managed a win at Road Atlanta. The Dodge Shelby Turbos of Team Shelby dominated the series.

In SCCA rally competition it was win some, lose some. Millen had lightened his four-wheel-drive RX-7 and revised its front suspension, and it paid off with an early lead in the series. But Buffum had traded his ex-works long wheelbase

The team of Johnson/Dunham/Terada won GTU honors for Mazda and Team Highball at the Daytona 24-Hour event in 1985. Mazda

Audi Quattro for a Group B Sport Quattro in mid-season, and after October's Press-On-Regardless Rally Millen and Buffum were tied at 105 points. Again at the finale, it was Buffum first, Millen second. Although Buffum took the season's driver's laurels, Mazda received consolation in winning the manufacturer's championship.

And finally—for something completely different—an RX-7 was entered in the Hong Kong/Beijing Rally. Pierre Honegger and Grant Wolfkill dodged Flying Pigeon bicycles to finish fifteenth out of thirty-six in the 2,000-mile epic.

Roger Mandeville, 1983 GTU champion, added the GTO driver's crown in 1984. Mazda

Chapter 9

1986
Finally, a New RX-7

The 1987 RX-7 Turbo debuted in March 1986.

Without a doubt, by the mid-1980s the RX-7 had grown stale. Not that it was anything less than the original introduced in 1978. Indeed, it was much improved, better in almost every way. It was faster, handled better, and was even more reliable—especially the 13B-powered RX-7 GSL-SE.

But the automotive landscape, indeed the mood of the country, had changed. Whereas 1979 had been bleak, with high inflation and higher interest rates, hostages in Iran and Joan Claybrook in NHTSA, in 1985 the country was in "the longest postwar period of growth and expansion" and the new NHTSA boss, Diane Steed, drove a Porsche. In 1979, the RX-7 had been a beacon in a world of emission controls-strangled automobiles. In 1985, there weren't enough magazine covers for all the hot new models, and the first generation RX-7 had been taken about as far as it could go. A new RX-7 was long overdue.

Of course the 1986 RX-7 had long been under development, in some ways

> *By the mid-1980s the RX-7 was faster, handled better, and was even more reliable. Still, a new RX-7 was long overdue.*

beginning just after the introduction of the original RX-7. Shortly thereafter, at Kenichi Yamamoto's direction, Chief Project Engineer Akio Uchiyama went to the United States for a three-month visit to learn how Americans—the new sports car's primary customers—actually lived. He visited RX-7 owner's homes, inspecting even basements and refrigerators! All in all he had more than 100 interviews and contacts. From this trip, aptly titled Operation Feedback, revisions to the original RX-7 were made while development of its replacement was deferred.

The new car project wouldn't actually begin for three years, until June 1981, a particularly long time for a Japanese manufacturer. The new project was designated P747, a name chosen by Akio Uchiyama because it gave no indication what the vehicle would be. If anything, it gave the impression of a racer, like Mazda's Le Mans Types 717C, 727C, 737C.

The project began with an evaluation of what the new car would be. Three concepts were considered, the Realistic Sports Car, the Technologically Advanced Sports Car, and the Civilized Sports Car.

The 1986 Mazda RX-7 (base shown) was literally new from the ground up. Critics, however, thought the new RX-7's profile too greatly resembled that of the Porsche 944—which actually had been Mazda engineers' performance bogey.

The first was an evolution of the original, with a live rear axle, disc/drum brakes and 13-inch wheels. The second was a no-holds-barred technofest with electronically controlled everything. The third would have four wheel discs and independent suspension with engines from 110hp to a turbo 180hp.

All three concepts were shown at product clinics in the United States in the spring of 1982. At a time when the median production price for an RX-7 fell just below $10,000, the concepts were priced at $9–12,000, $13-15,000, and $11-13,000 respectively.

The Realistic and Technically Advanced proposals were probably little more than straw men for the Civilized Sports Car favored by Uchiyama. The Realistic proposal would be hard pressed to meet current or coming competition. The TASC was just too great of a leap from the original car.

It might be a Civilized Sports Car in name, but Uchiyama wanted the emphasis on Sports. It would not be a Nissan ZX—despite that model's market success. It would not be, Uchiyama asserted, "a high-performance luxury car. It is and will remain a sports car of superior quality."

Uchiyama knew, however, that the United States was more than just an ocean away from Japan. His sojourn had taught him that it was indeed a different place, with different ideas and ways of using and even seeing the automobile. It was his experience that led a small group of young designers to move to the United States for a two-month crash course in the country and its culture. One exterior designer recalled day after day of sixteen-hour days, taking in shopping centers, sports car races, and (of course) Disneyland.

Fun but exhausting, it was the beginning of the design phase of P747. From temporary quarters in Mazda's Orange County, California, offices, the design team produced twenty sketches of the next-generation sports car. Ten were selected for consumer clinics in the United States in March 1982, from which two primary contenders were designated by Hiroshima. One, an evolution of the original RX-7, was chief project designer Yasuji Oda's favorite; the other was a more radical design.

Full-sized models were flown to California, and in October 1982, shown at consumer clinics alongside the Porsche 944, Toyota Supra, Nissan ZX, Corvette, and RX-7. Mazda took pains to hide the origin of its models, even allowing no Japanese on the test site (a challenge in California).

At the clinics, the revolutionary Mazda model scored better than the evolutionary, though both bracketed the Porsche's point total. The common criticism of the more radical P747 model was its stubby nose; back in Japan the nose would be lengthened, though Mazda designers insisted this was done for aerodynamics as much as the clinics.

In February 1983, the more radical design was chosen as the basis for P747. Running prototypes would travel to the United States and Europe for testing and evaluation, both mechanical and marketing. It was beginning to look like a winner.

What made a winner? The designers had a list. They wanted windshield rake for aerodynamics, opting for a steep 63.5 degrees—not quite the Alfa Romeo GTV's 65 degrees, but without the problems that slant would create (a vast dash, windshield reflections, and wiper storage, to name a few). Exterior designer Takashi Ono wanted a low "where's the engine?" hood profile, but the engine, small as the rotary is, required a hood line an inch higher than he would have liked. The fender line yielded to mechanical requirements as well, raised slightly to clear the top of the strut mounting. Another missed wish-list item was a "hidden" A-pillar, lost to sealing problems and cost.

The designer's successes were the short rear overhang, hidden drip rails, the look of big wheels and tires, and a front air dam moved forward to make the P747 look different from the original RX-7. The designers finally got their fastback backlight with the single-piece rear glass, though framed in steel. On the other hand, with two latches, the backlight contributed to chassis rigidity, and even without the backlight the P747 chassis was stiffer than X605's. The frame also allowed the glass to be thinner and therefore lighter, saving 10lb. The glass, however, was the largest in Japanese automaking history, requiring the supplier to buy a larger oven to make it.

Flush door handles were part of getting a drag coefficient of under 0.30. Actually, the base model would have a 0.31 Cd, though the optional aero kit—a soft-surfaced foam-filled urethane rear spoiler, front air dam extension and "deflectors" on the rocker panels—reduced the Cd to 0.29.

The designers also wanted a soft front end for better looks, aerodynamics and stone chip resistance. Mazda developed its own secret formula "R-RIM" (fiberglass-reinforced reaction injection molded) urethane, applying it prestretched to accommodate for the material's expansion when subjected to heat and humidity. Mazda prototypes tested in Puerto Rico in the summer of 1984 successfully survived 110 degrees Fahrenheit and 90 percent humidity without deformation or discoloration.

Suspension design was assigned to Jiro Maebayashi. While it was given that P747 would have MacPherson struts and an independent rear suspension, the only requirement for the latter was some form of toe control. Retaining the live axle was not an option for Mazda's late-1980s sports car. Ironically, the rear suspension

The RX-7 GXL was recognizable by its five-lug baloney-slicer wheels and "GXL" front fender decal. It featured electronically controlled shock absorbers to tailor ride and handling on the fly.

design work fell to a young engineer from Mazda's department primarily responsible for trucks. Yet Akio Uchiyama was so impressed with young Takao Kijima that he pressed him into service on P747, convincing his boss to set up the special project team for development of the sports car suspension within the truck chassis design department.

Double A-arm rear suspension was considered. Used on race cars, it could give a sports car exceptional handling, and toe-control could be easily added. But double A-arm suspension requires a lot of space, space needed by P747 for the rear seats of the 2+2 model to be offered in all markets. Racing suspensions also use solid ball joints at linkage points, and though good for precise control, they transmit too much noise and harshness for a road car. Semi-trailing-arm suspension was therefore selected, with Chapman struts as a backup.

Although Mazda had experimented with active four-wheel-steering in its MX-02 prototype/1983 Tokyo show car and would later offer it as an option on the 626 and MX-6, it was too far advanced for P747. Kijima instead proposed a kingpin axis in the rear hub: A lower ball joint and an upper cylindrical bushing and another bushing ahead of and below the wheel center would be "tuned" to provide desired toe changes. Extensive computer simulation suggested bushing location, hardness and compliance.

Two mules were built to test what was called the "KM hub"—named after Kijima and Maebayashi. One had a semi-trailing arm from the larger Cosmo model, the other a strut rear suspension with two lateral links and a single trailing link. Initial results were disappointing; although cornering was enhanced and trailing throttle oversteer—the bane of semi-trailing arms—was diminished, the car's handling felt sluggish.

Mazda racing ace Yoshimi Katayama suggested the problem lay not in the KM hub but rather in the semi-trailing arms, and recommended double A-arms be used. Kijima analyzed the problem as having bushings in the hubs that did just the opposite of what was needed—easily corrected without resorting to double A-arms. To Kijima the more difficult dilemma was the suspension's lack of anti-squat and anti-dive, caused by moving the semi-trailing arms' chassis pivot points downward 2.4mm to allow room for the rear seats of the plus-2 configuration. This put those points below the wheel axis, allowing uncontrolled squat and dive actions.

The answer was splitting the semi-trailing arm. The outer leg was unmodified, but the inner arm connected to the main arm and the chassis (at a point more to-

The RX-7 Sport was de jure *an option package but* de facto *a fourth model.*

ward the center of the chassis) by ball joints. This formed a wide-base trailing arm with strong lateral location. Camber change, more severe with a wide-based arm, would be controlled with another arm attached to the frame with a double-jointed link to clear the rear seat's low floor. A mule using this rear suspension and MacPherson strut front suspension was completed in May 1983 and tested that summer, along with a Porsche 944, at the Nurburgring in Germany. The car was much improved; in fact, the only handling-related incident occurred when Uchiyama crashed the 944, surprised by its lift-off oversteer after getting out of the Mazda mule.

The one shortcoming with the suspension, however, was its lack of "turn in"—it didn't start turns crisply. The prototype suspension was safe but dull and very unsports-car-like. This was because toe-in was induced by any lateral force. With the rear wheels immediately turning inward, the car resisted turning.

The solution came from the experimental active four-wheel-steering car. That car steered its rear wheels in the same direction as the front at high speeds for added stability. At low speeds it steered them in the opposite direction to the fronts for a smaller turning circle. Hirotaka Tachibana, however, was fooling with one of the cars that had a mode selector switch, and switched the mode to opposite steering while driving at a higher speed. Why not, he thought, calibrate initial opposite steering into the P747's KM hub? Kijima devised a firmer half-moon section to the front bushing that at low input kept the bushing from providing any toe-in. At higher lateral loadings, the bushing's resistance was overcome and a stabilizing toe-in effect took place. In high speed transients, the momentary lag before the toe-in effect could then take effect allowed the semi-trailing arm's toe-out to provide good initial turn in. Transition occurred around 0.4 to 0.5g.

While all this care was given to toe effects, rear wheel camber was maintained at -1.5 degrees. Leaning the tire inwards compensates for body roll in cornering and helps keep the tire's contact patch flat on the pavement.

Mazda called its rear suspension DTSS, for Dynamic Tracking Suspension System. The KM hub was also given its own acronym for public consumption: Triaxial Floating Hub. It would be used on P747.

If there was another problem, it was noise. To cure this, the rear suspension and final drive were mounted separately. A rear subframe was provided for suspension pickup points, the subframe then mounted to the unit body with rubber bushings of varying hardnesses selected to tune out the most offending vibration frequencies.

Forged-aluminum arms were selected over the cheaper and more obvious cast-aluminum arms; these aircraft-quality parts were made by Kobe Steel and required a new magnetic test facility for testing every arm for cracks; Mazda's need would be for 10,000 of the right and left interchangeable parts every month, far more than the typical aircraft contract.

While the rear suspension was all new, Kijima stayed with MacPherson struts up front. The strut itself was conventional, with an internal shock absorber and fitted with concentric coil springs, tapered at the bottom for tire and wheel clearance. The struts were strengthened by increasing the shaft diameter from the first generation RX-7.

The lower arm was an exquisite aluminum forging, an A-arm in function (eliminating the trailing link of the classic MacPherson suspension) with one arm extending inward from the wheel hub, the other extending towards the rear. Each A-arm was mounted to a front subframe also carrying the engine and rack-and-pinion steering. For anti-dive characteristics, the A-arms were mounted at one degree from horizontal, the rearward pivot higher than the front. Special bushings allowed horizontal movement of the A-arm, damping road shocks, while at the same time being very rigid laterally for precise location.

The front anti-roll bar was conventional, attached to the A-arm with links. Standard front anti-roll bar diameter was 22mm, changed to 24mm with a sports suspension that was standard with the Turbo and optional on an upgrade model. The rear anti-roll bar was 12mm standard, 14mm with the sports package.

A luxury, or perhaps better said Grand Touring, version of P747 would be equipped with a refinement of the automatically adjusting shock absorber system used on the Japan-market Cosmo and, in simplified form, optional on the 626. Called AAS, the system varied damping by rotating the shock absorber shaft. Three shock settings were provided, Normal, Firm, and Very Firm. A microprocessor was dedicated to analyzing vehicle dynamics to set the shocks. Sensors read steering angle (for roll control), brake fluid pressure (for controlling dive), and the accelerator (to reduce squat). It also read vehicle speed and the Normal/Sport

AAS Control Modes		
Normal Setting	Front shock	Rear shock
low speed zone	N	N
high speed zone	F	F
roll, squat, and dive	VF	VF
Sport Setting		
low speed zone	F	F
high speed zone	F	F

switch on the console. Thus by comparing steering angle and vehicle speed, the system could estimate g-force and set the shocks accordingly.

Standard on the luxury model and included with the Sports package (otherwise optional) was electronically-controlled power-assisted steering. Data on vehicle speed and hydraulic pressure were monitored constantly, and steering angle every 0.3sec. The steering system microprocessor used the steering angle and vehicle speed to compute lateral G-force while hydraulic pressure was plotted against a standard map. From this data the system adjusted steering assist, reducing it generally at higher speeds or higher lateral-G cornering for better steering feel, increasing it at lower speeds and tighter turning conditions for easier parking and low speed maneuvering.

Engine development proceeded during this time as well. Although three engines were initially proposed for P747, a carbureted six-port 12A was dropped in 1983. Progress on the fuel-injected 13B was yielding an engine that made more horsepower and beat the 12A's fuel mileage as well. This engine was first seen in the United States in the 1984 RX-7 GSL-SE (and in the Japan market Cosmo/Luce intermediates at the same time); the sophisticated airflow and fuel control were features of the 13B DEI EGI engine and allowed the production of a respectable 135bhp with a broad torque spread. Eleven more horsepower would be on tap as Mazda engineers discovered that the "tournament" effect—the shuttling of power pulses back and forth through the intake tracts—could be increased by reshaping the plenum. In the GSL-SE, the plenum had been box-shaped. For P747, the plenum would be curved to better sling the pulses from one side to the other.

Fuel injection was improved by going to dual fuel injectors. The GSL-SE had a single injector in the intake passage close to the intake port ("semi-direct" injection, in Mazdaspeak), the P747 engine added another injector in the secondary tract. Both injectors had "mixing plates," perforated flat surfaces against which fuel would be sprayed to increase atomization. Atomization was also increased by the use of a smaller injector in the primary tract. The fuel injection system, digitally-controlled Bosch L-Jetronic made by Nippon Denso, had timed pulse rather than the continuous injection of the GSL-SE's engine. The new 13B also got a bigger air cleaner, wider throttle valve and intake manifold runners and a new Denso pendulum airflow meter.

Yet another new apex seal was devised to replace the old two-piece design, this time a three-piece cast-iron-alloy seal with an upper and lower horizontal section with a wedged end for a snug fit. The new seal "swayed" back and forth less as pressure in adjoining chambers changed, and the thinner sections adhered more closely to trochoid irregularities. The seals, at 2mm wide instead of the earlier 3mm, also created less friction and wear. Side seals were thinner as well.

The ignition was still electronically controlled, although for improved fuel economy, the leading spark plug was moved back in the cycle (5mm in the trochoid housing) so that blow-by would be reduced as the rotor tip passes the spark plug hole. By firing later, there's less pressure in the chamber. Lubrication was improved by micro-channel porous chrome-plating, a technique already in use in the European 12A and Japanese turbocharged 12A engines. The process improves trochoid surface lubrication by interconnecting the pinpoint holes of the microporous plating system used with the earlier 13B.

Another improvement in lubrication

Although the production RX-7 Turbo would have a "Turbo II" decal, the "long lead" magazine test cars and this press release photo had a "GT" decal. Mazda

Cutaway drawing of 1987 RX-7 Turbo. Mazda

was the addition of oil injection into the primary intake manifold and directly into the trochoidal chamber (in carbureted engines, oil was injected into the fuel in the carburetor). Separate pumps fed the oil injection system and the eccentric shaft bearings. Oil was also sprayed against the inner surfaces of the rotor, the engine in effect cooled by oil as well as water. An oil cooler was installed just ahead of the water radiator. A thermostat shut off oil to the rotor to prevent overcooling, reducing cold-running fuel consumption by up to 12 percent.

On naturally aspirated engines, a multichamber exhaust port insert quieted the exhaust note. Exhaust emissions were controlled by a closed-loop feedback system with three-way catalytic converters. Two pre-converters preceded a main monolith converter with three distinct sections (two three-way and one oxidizing). The P747's dual tips were not simply for show, as the pipe from the converter was split into two pipes to two mufflers, each with ten liters capacity, compared to the single 17 liter muffler of the GSL-SE.

The naturally aspirated rotary reached new peaks of torque and horsepower in the second generation (1987 Sport shown).

transmission.

A second engine was in development at the same time, combining the turbocharger technology from the GT-X and the increased displacement of the GSL-SE, all to meet the horsepower goal of 170 set at the beginning of the project. Late in 1983, the ante was upped to 180bhp in response to the newly emerging horsepower race. But even with added displacement, conventional turbocharging hit a wall at about 160–165hp. Increased manifold pressure yielded not power but detonation. How could the engine meet a production schedule—final blueprints would be due soon—and the power goal?

Hiroshi Ohzeki was not called "Mazda's incumbent Mr. Rotary" for nothing. A variable-nozzle scroll turbo, as used and quickly abandoned by Chrysler, could have been employed but there were doubts about reliability and concerns about cost. Ohzeki instead devised the twin-scroll turbo which achieved a similar effect but with much less complexity. At low engine speeds exhaust gasses are directed through a narrow port in the turbo and directed at the turbine blades at almost a right angle, increasing the effect. At higher volumes (beginning at 2500rpm), a trap door (made of HK35 heat-resistant cast-steel alloy) in the turbo

So extensive were the changes to the 13B engine that it almost deserved a new name. Horsepower alone would have justified it. The 13B produced 146bhp at 6500rpm, with 138lb-ft of torque at 3500rpm. According to factory tests the 1986 RX-7 could turn 0–60mph in 8.0sec and do a 128mph at the top end. EPA fuel mileage figures were 17mpg city, 24mpg highway, with either manual or automatic

inlet opens and the exhaust is directed at the turbine at a more "normal" angle. Hitachi supplied a new turbine blade shape for optimum effect. A normal exhaust wastegate was used to regulate maximum boost, set at 6.2psi (0.42 bar) at 2000–5000rpm, decreasing to 5.4psi (0.42 bar) by 6000rpm. Because the turbo rotary used more air than any airflow meter in Nippon Denso's catalog could handle, a new meter had to be developed that could manage the widely varying demands of the engine. A special heavy-duty two-speed fuel pump was required as well.

An air-to-air intercooler was mounted above the engine, the most obvious feature of the underhood landscape, to feed cooling air from a functional hood scoop. The intercooler reduced the temperature of the intake charge, heated when compressed by the turbocharger, for improved volumetric efficiency. While the Japanese 12A turbo used signals from the ignition pulse, intake temperature, and boost sensors to detect the "knock-prone zone" in which ignition would be retarded and fuel injection enriched, the 13B turbo used a piezoelectric knock sensor to "hear" incipient knock, retarding ignition via electronic controls.

Although the Dynamic Effect Intake system was used with the turbocharger—and because the twin-scroll design worked well despite its large intake volume—the Turbo II engine, as it would be known commercially, had only two ports

The interior of the 1987 RX-7 Turbo, like other second-generation models, had a two-spoke steering wheel and centrally located tachometer.

per chamber (primary and secondary). The multichamber exhaust port insert of the naturally aspirated P747 engine was replaced by a single chamber insert, giving the exhaust a straighter shot at the turbocharger while insulating the exhaust to keep it hotter for the turbo and the catalysts. For maximum effect and minimal lag, the turbocharger was tucked as close to the engine as possible. The turbo engine's compression ratio was dropped to 8.5:1 from the naturally aspirated engine's 9.4:1.

Unique to the Turbo was the application of a Teflon coating to the trochoid working surface. The coating would be scraped and burned away, but not before assisting the bedding-in of the trochoid and apex seals. And should the supply of lubricant be interrupted, a small amount of Teflon squeezed into the trochoid surface's micro-channel pores would reemerge and provide temporary protection against damage.

While all Mazda rotaries since 1974 had an inner core of sheet steel (called SIP, for sheet-metal insert process) to provide a more durable surface on which to apply the chrome plating, Mazda substituted a stronger material (HPC-50 instead of HPC-45) for the Turbo. Mazda also reinforced the rotor-gear assembly with twelve instead of nine spring-loaded pins securing it to the rotor body. Not only was this stronger but it also reduced the possibility of vibration damage.

The Turbo II engine met output goals with ponies to spare: 182bhp at 6500rpm and 183lb-ft at 3500rpm. Despite an EPA city rating of 17mpg, heavy-footed *Motor Trend* testers were only able to match that on the highway. Others could not even match it: *AutoWeek* chalked up a cumulative week's worth of driving at 11.8mpg. Undoubtedly that was with every staffer pumping as much high-test unleaded through the injectors as possible. Although *AutoWeek*'s result probably was not representative of real world use, it was possible to use prodigious amounts of fuel.

Two transmissions choices were offered in P747, a five-speed manual and a four speed automatic. The naturally aspirated engines would be backed up by the Type M five-speed, as used in the GSL-SE, but with a larger synchronizer cone (65mm vs. 61mm) to improve shifts between first and second gear. The shift lever was also moved about 1in rearward, allowing the shift lever to be more vertical. The lever diameter was increased for a sturdier feel.

The Turbo's standard five-speed was

The turbocharged engine produced 182bhp, the highest ever for a production rotary.

not the same transmission as used with the standard engine, because the Turbo engine produced some 30 percent more torque. Mazda brought back the Type R gearbox, a sturdy design that had been used with the RX-3, though now with high-mesh gears (these had finer teeth, more contact area, and produced less noise). First through third gears were specially machined and welded for an extra 3mm gear width. Larger synchronizers were fitted between first and second and second and third. The Type R transmission also differed from the Type M by having a separate, instead of integral, bellhousing.

Both the Turbo and naturally aspirated engines got larger clutch discs. The lower-horsepower rotary got a clutch with outer and inner diameters measuring 8.86in and 5.9in respectively, while the Turbo's clutch disc measured 9.4in and 6.3in, outer and inner.

The optional automatic was a JATCO (Japan Automatic Transmission Company) L4N71B, a four-speed with a converter lockup clutch. Lock-up was set to engage in fourth gear only above 44mph, remaining disengaged for smoother operation in specified conditions. Shifts into second and third gear were also smoother compared to earlier versions. For added reliability, there was an oil cooler in the lower header of the radiator, special ducting for the torque converter housing, and impeller cooling fins.

While the naturally aspirated cars had a conventional driveshaft, the Turbos had an inner and outer sleeved shaft with torque transmitted by three pressed-in and bonded rubber rings. The rubber damped high-speed resonance noises. A multidisc limited-slip differential was standard on all but base models.

Befitting a Mazda sports car of the late-1980s, brakes were four-wheel disc, although premium models got better components. Base models had single-piston floating-calipers on 9.8in ventilated discs at the front and solid 10.3in discs at the rear. The Turbo and the upgrade model had four-piston aluminum (instead of cast iron) calipers at the front, with single-piston floating-calipers at the rear. The aluminum caliper weighed 7.1lb, compared to 9.5lb for the base model's cast iron caliper and 12lb pounds for the caliper of the first generation RX-7. All discs for the Turbo's system were ventilated measuring 10.9in and 10.7in diameter front and rear, respectively. The bigger brakes also required a bigger vacuum servo (9in versus 8in). Necessity forced one unplanned benefit on Mazda: the new rear suspension system meant the rear caliper had to mount on the disc's front side. Because it was closer to the semi-trailing arm pivot, it had less unsprung weight effect.

Antilock braking was not available at model introduction but was offered shortly thereafter. It was based on Bosch's ABS but with a Mazda-developed accumulator circuit to reduce pedal pulse.

Wheels provide long distance identification of the three versions of P747 sold in the United States. The base model had 14x5-1/2in alloy wheels, recognizable by their four lugs and eight circular holes in the disc, mounted with 185/70HR14 tires. The luxury upgrade wore 15x6in wheels mounted with Bridgestone Potenza RE71 205/60VR15 tires. These wheels had five lugs and five angled spokes. The Turbo would complete the progression with 16x7in rims and Goodyear Eagle GT 205/55VR-16 tires. These wheels had five lugs and ten turbine-like vanes around the perimeter.

The interior of the original RX-7 would have looked rather plain and simple by mid-1980s standard. Such could not be allowed for P747. Nor could any fads that could quickly fall from fashion. Like the exterior, interior design began on paper at Mazda's U.S. headquarters in Irvine, California. The original designers, one American and three Japanese, were joined by Yasuo Aoyagi who was charged with responsibility for coordinating the interior design.

The interior designers began with a "bathtub" concept, a tub being "that most secure and personal of spaces." The designers started with single king-size and twin tubs, working these into automotive interiors. Instrumentation ran the gamut from conventional needle-and-dial analog to electronic with digital readouts, electronic graphics, and a cathode-ray tube. Fortunately, by a nine to one vote, the more traditional approach was followed; the only digital displays were on the clock and some optional radios. The final instrument panel had red lettering with orange needles (to the misfortune of the color-blind) and centered the 4-1/2in, 8000rpm tachometer. To the tach's right was a slightly smaller speedometer marked to 150mph. To the tachometer's left were four smaller gauges with ninety degree sweeps for oil pressure, coolant temperature, fuel level and battery voltage, the last replaced by a boost gauge in the Turbo.

While Mazda designers were wary of off-beat and unergonomic controls, they believed that something unusual, though functional, was required for the revolutionary P747. For this they followed the 626, placing controls on the sides of the large instrument pod. Vision controls—lights and such—were placed on the left side (although the emergency flasher switch was placed within reach of the passenger) and wiper and washer switches placed on the right. Most controversial were the "flipper switches" on the instrument panel, particularly the unconventional turn signal lever.

Mazda hoped to widen the appeal of its new sports car by offering it as a two-seater and as a two-plus-two in all markets. But between the two versions there were no chassis differences, so the rear seats weren't really suited for full-size adults—in fact, the seats were not much bigger than the storage bins they replaced. The seatbacks in the 2+2 folded for maximum storage space.

Interior features standard on all models included remote hatchback and fuel lid releases, cargo area and glovebox lamps, luggage hold-down straps, cloth upholstery, carpeting, and a clock. The luxury model added a leather-wrapped tilt steering wheel, velour upholstery, cruise control, power windows, locking storage boxes, and air conditioning; leather seating and a security system were optional.

The RX-7 was scheduled to appear as a 1986 model. That meant cars would have to be on the Hiroshima docks in the summer of 1985—preproduction would have to be off the line even sooner.

Final project approval for the 1986 RX-7 was originally scheduled for June 1983. With one month to go, interior and exterior design had been set and mechanical specifications nailed down. But as the sum of the parts were totaled, P747 had some too much. It was overweight. The whole project might be abandoned. The problem was that P747, to meet fuel economy targets, would have to fit in the U.S. Environmental Protection Agency's 2,875lb class, or the 3,000lb class with the automatic transmission. If these targets were not reached, P747 might not achieve the minimum combined city-highway rating of 22.5 mpg necessary to avoid the federal Gas Guzzler Tax. (Fuel economy ratings, as calculated by the EPA, are not recorded by driving a car on the open road or even on a test track. Instead, a car is run on a chassis dynamometer, a set of rollers with resistance calibrated to simulate rolling and wind resistance. The resistance is set by weight classes, therefore the necessity to stay under certain breakpoints.) Not only was there a need, from a marketing standpoint, to show the best fuel mileage possible—admittedly less critical for a sports than economy car—there was also the matter of Mazda's pride. As Uchiyama would later say, in reference to the early to mid-1970s, "Once was enough. Never again will we be called a guzzler."

Weight could be reduced by scrapping the independent rear suspension; a live axle would be lighter overall. A three piece rear window and 13in wheels would save weight, too. Or there was the zero option: cut losses, drop P747 and continue building the fully amortized X605. Uchiyama would later admit that P747's status was perilous, only underscored by the product approval date delayed to November. That could be the kindness that killed P747.

Desperate times call for drastic measures. Top management had ordered a freeze on development for one month so that weight and cost could be examined as well as whether the project should even move forward. In a gutsy move, senior managing director Takushi Mitsunari, without waiting for concurrence of the other directors, authorized the ordering of production dies and presses during the freeze. This was necessary if production deadlines for the 1986 model introduction were to be met. At stake was around $80 million, the cost of the order.

Uchiyama then had a bit of breathing room but also an obligation to Mitsunari. He launched "Operation Gram Per Head," in which every designer and engineer working on P747 was asked to devise a way to save one gram from the weight of the car. The staff dismantled a prototype—by this time complete cars were running and being evaluated—and spread the parts on rows of mats on the floor of the design center's auditorium for the designers and engineers to pick and choose for weight reduction. It was in OPGH that fabricated steel A-arms became elegant forged aluminum, saving 4.4lb each. Alloy wheel hubs front and rear saved another 14.7lb per car compared to cast-iron hubs. An alloy engine bracket pared 2.4lbs. An aluminum final drive rear cover and mounting bracket reduced vehicle weight another 4.4lb.

Every department involved in the project participated: the engine group saved 3lb in the rotors alone, reducing the weight of the car and serendipitously reducing frictional losses and engine inertia.

Weight sensitive models were fitted with aluminum hoods to keep them under the limit. Space-saver spare tires were mounted on neat aluminum wheels as another weight-saving trick. Even more exotic was the aluminum jack imported from Germany to save 3lb on every P747 made.

The operation was more than successful. Even loaded with accessories P747 weighed in at 2,626lb curb weight. That even compares well against the 2,500lb pound 1985 GSL-SE, a much less "technologically dense" car. P747 also would have better weight distribution, near 50/50 with two up and a full tank of gas, compared to the GSL-SE's 54/46. But most important, at least from the standpoint of the Gas Guzzler tax, was the EPA fuel mileage rating: 17mpg city and 24mpg highway. That was better than the GSL-SE, which barely cleared the Gas Guzzler 21.5mpg minimum for 1985.

It was not without cost, however. More alloy components would go into P747 than any Mazda since its first car, the tiny 1960 R360 coupe. And light alloy costs more than steel. The aluminum front A-arms were more than double the cost of fabricated steel arms. On the other hand, had not the weight bogeys been met, P747 might never have seen production. In a peculiar twist, the federal law enforcing fuel economy made P747 a better sports car, lighter, faster, and better handling.

Starting with the RX-2, Toyo Kogyo had followed a simple process for naming its rotary-powered automobiles: just add a number. The progression led up through the RX-5 (sold in the U.S. as the Cosmo) and then skipped to the RX-7. There was no RX-6, it's said, because "x" and "six" are difficult for the Japanese to pronounce (although this doesn't explain the 626 or MX-6). Because P747 was virtually an all new car, RX-8 would have been the logical name. Even skipping all the way to RX-11 was suggested—someone must have thought it would be lucky. Ultimately, it was decided to continue with RX-7, a name well known particularly in the all-important North American market. Finally there was the argument that "if the car is good, the name will be good."

Mazda claimed the Sport's petite rear spoiler was more effective than a larger one at cutting drag and lift.

Who, after all, would name a car Porsche or Buick?

So P747 became the RX-7, second generation. The base model received no special designation and was simply called the RX-7. The "high content" luxury/ grand touring was named the GXL. The turbocharged model, appearing in March at the Chicago Auto Show as a 1987 model, was called the Turbo—with "Turbo II" decals on its front fenders. This mystified some, and even Mazda's United States representatives couldn't explain the "II" (it was likely used because it was the second turbocharged RX-7, if the first brought Stateside).

Further confusing the issue, was the initial badging of the Turbo with "GT" decals on its front fenders. This apparently was done only on the "long lead" car used for early tests by the monthly magazines.

If any Americans were confused, it sure didn't stop them from buying RX-7s. Sales reached a new all-time high: 56,243, out of 72,760 worldwide, for calendar 1986.

1987

Following the introduction of the all-new RX-7 in 1986, one might not expect much new on the 1987 RX-7. Indeed, that was the case. The model lines were carried over unchanged except that anti-lock braking was added as a separate $1,300 option on the Turbo and GXL. Another change was the addition of a starter interlock to the clutch. This prevented starting unless the clutch pedal was depressed. And finally, the Turbo engine was given what one reviewer called "a dose of refinement" so that it no longer felt "like an aftermarket conversion."

RX-7 sales declined in a normal second-year pattern, totaling a still strong 38,345. Mazda's increasing strength was demonstrated in the company's purchase of Ford Motor Company's assembly plant in Flat Rock, Michigan, which would be used to build the Mazda 626, MX-6, and Ford Probe. And "Mr. Mazda," Kenichi Yamamoto, was appointed chairman, Mazda Motor Corporation's top spot.

Chapter 10

RX-7 Competition 1986–1988
A New Era

Eight thousand five hundred and fifty revolutions per minute. Two hundred and thirty-eight miles per hour. Just as Racing Beat had taken the 1979 RX-7 to the salt flats and Don Sherman had driven it to a new class speed record, the same team would match up for another assault on the salt in a second-generation RX-7. And another record would fall.

For 1986 the class of choice was Grand Touring Sports. It allows substantial chassis and engine modifications but requires the body to be stock. The results are cars that look fairly normal, if one disregards the skinny tires, the belly-to-the-salt stance, and if one doesn't look inside or—especially—under the hood.

The project was two years in development. Mazda supplied the turbo rotary technology, and Racing Beat prepared the engines and set up the car. The car,

Racing Beat and Don Sherman teamed up for another speed record based on the 1986 RX-7. Racing Beat

The second-generation RX-7 drew first blood on the salt flats of Bonneville, and the 1986 IMSA season saw the by then customary Mazda massacre in the GTU class.

per Sherman's description, was a race car under a stock body shell. To lower the car, few stock front-suspension parts remained, and the whole rear suspension was scrapped in favor of a simple five-link-per-side arrangement using a Porsche 911 stub-axle carrier and no brakes.

Mazda had originally designed the engine for GTP racing, but it had proven short on durability. Since long haul reliability was not an issue for Bonneville blasts, the engines were shipped to Racing Beat for development. The engines were side bridge-ported, but side porting was the only similarity to the stock turbo rotary. No fancy electronic fuel injection here, just Bosch Kugelfisher mechanical injection.

Two turbochargers were fitted, made by Hitachi of special alloy steel that would survive 2,100deg-F exhaust temperatures. And to help the intake charge keep its cool, an air-to-water intercooler was installed.

The intercooler was described as merely "medium-sized," but with water

Tom Kendall, shown here winning at Elkhart Lake in 1987, won the GTU driver's championship with the very same car with which Jim Downing took the title in 1982 and Jack Baldwin in 1984 and 1985. Mazda

The engine of the 1986 speed record RX-7 resembled the production version in concept only. Racing Beat

at 32deg-F from an ice-filled tank at the rear (coolant circulated by an electric pump), air leaving the turbo compressor at 280deg-F was chilled to 45deg-F on the cool side. The ice tank also helped by putting weight over the drive wheels. With 14.7psi boost (1 bar) and a compression ratio of 7.5:1, the double pumper 13B cranked out a prodigious 530hp at 8500rpm.

A dry-sumped Weismann Transmissions five-speed crash box (no synchronizers) sent the power back to a specially constructed locked rear end, with a final drive ratio of 2.00:1.

Rules require a two-way run to set a speed record, and although an all-stops-out return run topped out at 8550rpm—or 244mph—the average speed for both directions was 238.442mph, smashing the 201mph class record previously set by a Pantera in 1981.

While the second-generation RX-7 drew first blood on the salt flats, the 1986 IMSA season saw the by-then customary Mazda massacre in GTU. Amos Johnson started it off with a win at Daytona, but the dominant car would be the RX-7 driven by Tom Kendall, driving his first full season as a professional racer. The car was, no surprise, the same one that Jack Baldwin and Jim Downing had driven for GTU championships.

Ira Young had disbanded his Malibu Grand Prix team at the end of the 1985 season, and Clayton Cunningham bought the well-tested RX-7. Prepping and driver selection weren't completed until after Daytona and Sebring, however, and then the nod for driving duties went to UCLA student Kendall. Despite Kendall's inexperience, his steady driving yielded eight second-place finishes with a mere four class wins. That was good enough for the championship, however, and Kendall tallied sufficient points to finesse the title from Roger Mandeville, who had eight wins in his former GTO-class RX-7 downsized to run GTU.

Mandeville, though, had been dividing his time between racing and prepping a three-rotor RX-7 for GTO. The 450hp triple would still be horsepower short compared to the V-8s and turbos, but rotary reliability had worked before and at least the gap was no greater than it had been previous seasons. Debuting at the Daytona season finale, Mandeville finished eighth and confirmed Mazda's GTO underdog status for the upcoming 1987 season.

John Hogdal, however, played overdog at the SCCA runoffs at Atlanta. The sandbagging 1985 champ was gridded alongside RX-7 veteran and pole-sitter John Finger, leading him into turn one after the green flag fell. Finger led briefly, but with plans to move to GTU in 1987, Hogdal put his foot down and began to stretch out a lead before a detour into the pits. Finger finished third but his RX-7 was disqualified, along with the erstwhile sixth-place RX-7 of Michael Green, both for a "rear window discrepancy."

Dr. Bill Schmid gridded and finished third in his RX-7 in GT-3 behind Bruce Short, whose RX-3 nipped Schmid.

Off-road, or at least off-pavement, Rod Millen along with designer David Bruns prepared a new lightweight four-wheel-drive RX-7 with 1986-look bodywork and a peripheral-port 13B engine. But Buffum's turbocharged Audi Sport Quattro still had enough of an edge to make Millen a consistent second-place finisher. Millen threw in the towel and switched to a Group-A-legal Mazda 323GTX for the Olympus rally near the end of the season and won his class.

With the 323, Millen would finally have the success which had always been just 100hp out of reach with the RX-7. It was, however, the end of an era for Millen and Mazda's rotary sports car.

1987

IMSA's 1987 GTU season was almost a rerun of the 1986. Just as he had the year before (and the year before that), Amos Johnson won the 24-hour season opener at Daytona, and in the same car for that matter. It looked like it might finally be Johnson's year, too, with enough wins and good finishes for a lead in the points standings. But steady Tom Kendall was right there in the venerable ex-Downing/ex-Baldwin racer, winning when Johnson finished thirteenth at Lime Rock, and taking the points lead. Johnson won next at Watkins Glen, but Kendall's second clinched the driver's championship.

Mazda's manufacturer's championship in GTU was also aided by Al Bacon's victory—his first in six years of IMSA competition—and former SCCA champ John Finger's first pro win at Miami in his GT-2 RX-7.

Winning was not so easy in GTO. Mandeville returned with the three-rotor RX-7 to race factory teams from Ford, Chevrolet, and Toyota. The result was disappointment. Mandeville never finished better than fifth, and placed tenth in the championship standings. It was just not enough motor: the naturally aspirated rotary was short horsepower against the turbos and big V-8s.

John Finger backed up his Miami IMSA victory with a solid second place at the SCCA runoffs. Finger had the pole for the GT-2 race and held onto the lead for seven of the eighteen laps before letting the 280ZX of Morris Clement slip by. The highest placed RX-7 in GT-3 was Bill Reid in seventh place, who was also the top rookie in the class.

Ironically, Rod Millen built a specially prepared lightweight RX-7 for Pikes Peak, but it was Robby Unser in Millen's Group A 323GTX who won the first place class trophy. Millen's mount was his old 4wd rally RX-7 with a turbocharged peripheral-port 13B and wings and spoilers *a la* his Audi Sport Quattro competition. But Millen chose the wrong tires and could muster no better than sixth with his shredded treads.

1988

Tom Kendall repeated in 1988 as GTU champion, but the bad news was that he wasn't in an RX-7. Given an offer he couldn't refuse, Kendall switched to a factory-supported Chevrolet Beretta. And although Amos Johnson won the Daytona 24-Hour and Sebring, these would be the only GTU race wins for Mazda. Chuck Kendall, Tom's father, bought Tom's RX-7, and son Bart drove it to finish fourth in the driver's standings.

GTO was no easier for Roger Mandeville in 1988 than it had been the year before, but at least this year he had company. Clayton Cunningham changed his newer GTU RX-7 (which he campaigned as a rent-a-racer in 1987) into a GTO by installing a three-rotor motor and had an all-new three-rotor car ready by mid-season. With a variety of drivers through the season, the team would garner no better than a fifth-place finish.

At the SCCA's runoffs at Road Atlanta, John Finger's RX-7 was the fastest Mazda in GT-2, but was relegated to a fourth-place finish by a pair of Porsches and a quick Celica. And though Bill Golden placed third in GT-3 in his RX-3, the best placed RX-7 was Jeffrey Scoville in sixth.

Elsewhere for Mazda, Millen won a national rally championship, but in a 323GTX, while Richard Kelsey won the Rally Production GT class at Pikes Peak. For the RX-7 it was not a vintage year, but corporate changes were about to change Mazda-racing-as-it-had-always-been.

The Johnson/Dunham/Shaw GTU RX-7, on its way to winning the 1986 Daytona opener, precedes a Mazda-powered Tiga Camel Lights on the banking. AutoWeek file photo

The Clayton Cunningham Racing three-rotor GTO RX-7 was driven during the 1988 season by John Morton, Parnelli Jones, and P.J. Jones, among others, but never finished higher than fifth in class. AutoWeek/Richard Dole

Chapter 11

1988
Diversification

If 1987 had been a carryover year for the RX-7, 1988 held a bonanza of change. The GXL and Turbo returned, but the base model would now be called the SE, a new GTU model was added, and the RX-7 convertible (which had debuted in Japan in 1987) was introduced in the U.S. A special model for the 1988 model year was the Tenth Anniversary RX-7—celebrating, appropriately enough, the ten years since the introduction of Mazda's sports car.

The convertible was a landmark in its own way. Not only was it a new version of the second-generation RX-7, but it was also the first factory-built RX-7 convertible. It was truly a factory job, beginning life on its own assembly line in the Hiroshima factory where the basic assembly was welded up. It then joined the standard assembly line, along with a mix of other Mazdas, including RX-7 coupes, 929s, and 323s.

Extreme accuracy was the standard for the RX-7 convertible, the windshield header was permitted a mere 1mm of tolerance, compared to 5mm allowed for the coupe. The accurate assembly meant the top required no adjustment once installed, unlike other convertibles which often require extensive fitting. Because the top takes longer to install than most other operations, the RX-7 convertible was preceded and followed down the production line by a Familia/323, which required the fewest production tasks.

Extreme accuracy was the standard for the RX-7 convertible, the windshield header was permitted a mere 1mm of tolerance.

Despite the integrated roofline of the steel-roofed RX-7, the RX-7 convertible looked good enough to have been designed with the fabric top as the primary version. Not only was it attractive, it also had a drag coefficient of 0.33 with the top raised and 0.38 with it down.

Nor was Mazda content to make merely a simple ragtop; the top was a unique three-position affair. In addition to the usual up and down positions, the Mazda top could be run with the top "half up." Only the rear section of the top was soft fabric, while the overhead section was rigid and removable. Although it was simply called "the convertible" in the U.S., the Japanese market name was really more ap-

The Tenth Anniversary RX-7 was finished in monochromatic white.

The U.S.-market RX-7 convertible went into production as a 1988 model. With the top raised, the convertible was snug, aerodynamic, and even handsome.

Convertibles were finished on the same production line as other RX-7s. Mazda

propriate: Cabriolet.

Mazda seemed more proud, however, of the convertible's patented Windblocker. This was the brainchild of Takaharu Kobayakawa, then chief engineer of the RX-7 range. When he had earlier been on the 323 Cabriolet project, he insisted on testing the car in a rather hostile environment: in the mountains on a skiing trip—with the top down (yes, in Japan, which has hosted the Winter Olympics). He and his travelmates were, not surprisingly, frozen by the experience. But their shivering was not so much caused by air from the front as from the blast over the windshield which curled forward toward the passengers from behind. "Koby" found that a shield could be devised to stop that reverse flow, and that idea was used for the RX-7's Windblocker which could be raised for relatively draft-free motoring. With the windows up and the high-output heater on, convertible fans could go topless for much more of the year.

While the Japanese-market Cabriolet came only with the turbocharged 13B engine rated at 185bhp JIS and a choice of five-speed manual or four-speed automatic transmissions, Americans would get the 146bhp naturally aspirated rotary and the five-speed only. Why? The turbo and automatic would raise the RX-7's weight up into Gas Guzzler Tax territory. On the other hand, the Japanese buyer got the GXL's five-bladed aluminum wheels, while American customers found their convertibles equipped with 6.5in-wide BBS multispoke wheels.

The top included a glass rear window with defogging heat elements. Power-operated, the top required only the release of two windshield header clips and a twist of a dash-mounted control to lower the top. The rigid front panel had a lock that had to be released before the top was fully lowered, and a vinyl boot snapped into place for a finished look. For safety, the handbrake needed to be set before the top would raise or retract.

The convertible's body reinforcements worked well; *Road and Track*, for example, reported that on "particularly rough roads" the convertible exhibited "some mild cowl shake," though within acceptable limits, and gave Mazda "excellent marks for a rattle-and-creak-free interior, regardless of how brutal the road surface might be." The magazine lamented, however, the loss in performance caused by the convertible's extra weight and the taller final drive ratio—3.91:1 instead of 4.10:1—installed in the convertible to boost fuel economy. Acceleration from 0 to 60mph now required an additional 1.2sec. The only other complaint was the typical loss of rear vision with the convertible top raised.

Joining the convertible as a new RX-7 model for 1988 was the SE. This base model, previously just called the RX-7, was

The RX-7 Turbo was "refined" for 1988. The rear "wing" was standard.

The base model RX-7 was called the "SE" for 1988 only. Note "old style" side mirrors, finished in matte black.

Special fender badges commemorated ten years of RX-7 production.

identified by a decal on its front fender but was otherwise essentially unchanged from its predecessor. The standard anti-theft system, previously standard on the GXL, was extended to all models for 1988; the 2+2 version was available in SE or GXL trim.

The GXL and Turbo were back mostly unchanged, but Mazda brought a pair of new versions to market for 1988. One was the GTU, named in honor of the exciting monotony with which the RX-7 won the IMSA GTU class championship year after year.

The GTU was, if not a Turbo without a turbo, a more serious sports car than the base SE or luxury GXL. The GTU was fitted with the GXL's five-spoke 15in wheels with 205/60VR-15 Bridgestone Potenza RE71 tires and the larger RX-7 front brakes and ventilated rear discs, though neither ABS nor the GXL's electronically controlled shocks were available. The GTU also had standard vehicle-speed-sensitive power steering. Other additions were a "sport-tuned" suspension (same diameter anti-roll bars as the Turbo and GXL, firmer shocks and springs, and higher-durometer bushings), sport seats, aerodynamic body pieces (front air dam, rocker panel extensions, and rear spoiler), limited slip differential, and body-colored electric outside mirrors. A fender decal identified the GTU, which came close to Turbo handling without the added expense of the turbocharged engine, or its higher insurance premium.

The other new 1988 model was a special version of the Turbo. Honoring ten years of RX-7 production (beginning in 1978), the Tenth Anniversary model came in any color as long as it was monochromatic white. Even the wheels, 16x7in alloy in the same pattern as on other Turbo models, were white. But the Tenth Anniversary was a luxury hot rod, fully optioned with ABS, leather upholstery, and a special sound system. If the blindingly white finish did not ID the car, a Tenth Anniversary badge on each front

The Anniversary RX-7 had leather interior with a specially embossed steering wheel.

fender notified onlookers of its special status. The driver was reminded by an embossed steering wheel hub and special Tenth Anniversary floor mats.

Road & Track was critical of the Tenth Anniversary RX-7, finding it too hard for a luxury GT but too soft and too fully accessorized for a serious sports car. The turbocharged rotary was faulted for lacking top-end boost, with excessive lag and not enough additional power to make it all worthwhile. The magazine relented somewhat, however, saying the turbo motor had more torque than a Porsche 944, but added that was "damning praise." Editorial preference, between the Tenth Anniversary Turbo and the GTU, was for the latter which, the editors concluded, was "the *true* Tenth-anniversary version of the RX-7, and a worthy one."

With the U.S. introduction of the MX-6 sports coupe and the 929 luxury sedan, 1988 was a banner year for Mazda. A new overall sales record was set with 256,050 cars sold in the U.S. for the calendar year. But despite all the new versions of the RX-7, sales of the sports car slipped to 27,814.

Chapter 12

1989–1992
The Major Minor Change

It doesn't matter what the model is or who the maker is, the common knowledge of automobile marketing is that sales will peak in a model's first year, then taper off until a revision is brought out. It's partly what-have-you-done-for-me-lately and partly other makers releasing new and improved models. Manufacturers counteract the trend with special editions, minor revisions, and facelifts, as well as genuine additions and improvements. It's harder to hit a moving target.

So for 1989, Mazda reshuffled the RX-7 model line. The SE was gone, replaced by the GTU—but not the 1988-style GTU. Rather, it was the SE simply renamed GTU, the biggest change being an upgrade to 15in wheels. A new model, the GTU^S, replaced the GTU as the nonturbo ultrasports version, with specifications even more near those of the

Mazda built 100 1990 GTU^S models before discontinuing production (fog lights are owner-installed).

Underhood changes for 1989 were more than cosmetic. The rotary engine was revised, with more power and refinement for both the turbocharged and naturally aspirated versions.

Turbo. The convertible and GXL were back for their respective niches. And for the 1989 model year, the 2+2 was available in GXL trim only.

Mazda also gave the RX-7 a mild facelift for 1989. The most evident change was a body-color rub strip. Look closer and the nose cap was reshaped, with "Mazda" embossed (instead of a decal), larger "brake ducts," and integrated fog lights for the GXL, Turbo, and convertible, flanking a revised radiator opening. The rear spoiler, for those models so equipped, was reshaped, and new taillights, with circular lamps appearing under a smoked lens, highlighted the rear.

Underhood changes for 1989 were more than cosmetic, however. The rotary engine was revised, with more power and refinement for both the turbocharged and naturally aspirated versions. More compression, lighter rotors, and improved breathing raised horsepower and torque ratings and also improved EPA fuel mileage ratings. Side seal thickness was re-

The 1989 GXL had new wheels and body-colored side moulding. Mazda

duced from 1.0mm to 0.7mm, and 5 percent lighter rotors also gave the naturally aspirated engines an incredible 8000rpm maximum. By fully machining the combustion recesses, the compression ratios of each of the rotor's combustion chambers could be more closely matched.

The standard RX-7 engine received more horsepower and torque from the "variable dynamic effect" intake system.

Specifically, the engine modifications raised output of the naturally aspirated 13B to 160hp at 7000rpm, but more importantly, the torque curve was broadened. Torque went up by only 2lb-ft (to 140lb-ft) but was beefed up on the lower end by the operation of a rotary valve between the intake tracts. Opening at about 4500rpm, the valve enhanced the DEI (Dynamic Effect Intake) by altering the length of the path between intake ports, and thus changing the tuning of the pulses that provide the ram effect to the intake charge.

The airflow meter was changed to a linear type for reduced air resistance and more precise measurement. Instead of the conventional flap-type meter, the "linearic" meter used a sliding cone to measure air flow. The engine control computer speed was doubled and operation upgraded for faster response to throttle inputs. Even the oil pump was computer-controlled for more effective lubrication and reduced oil consumption.

For the Turbo, a new turbocharger, dubbed Completely Independent Twin-Scroll, fed both exhaust ports directly at the impeller. "This permits," explained Mazda, "the powerful exhaust pulse generated by the opening of the exhaust port to directly hit the turbine blades without interference from the exhaust of the other rotor." Combined with the twin-scroll design of the earlier 13B-based turbo-rotaries, it reduced turbo lag and increased torque. Electronic control of the turbo wastegate, which allowed a more closely calibrated response than purely mechanical control, also improved torque and response, providing peak torque from 2000 to 5000rpm.

Shifting was enhanced by a revised shift rod mechanism and a shorter shift rod, topped by a "gun-grip"-type knob, which was leather-covered in all but the GTU. Leather also covered the steering wheel and handbrake handle, except in the GTU. Automatics benefited from a new electronically controlled four-speed automatic with lock-up torque converter and "Hold" mode that allowed the transmission to be shifted manually in the lower three gears. The automatic was available in the convertible for the first time and was also optional in the GTU and GXL. The Turbo and GTU[s] required the five-speed manual.

Beginning in 1989, all RX-7s, except the convertible, had "mouse-type" passive shoulder belts. Note the three-spoke wheel (1990 GTU^S shown).

New-style taillamps (shown on 1990 GTU^S) had dual circular brake/tail lamps and single circular amber turn lamps under a smoked cover.

Other changes to the interior included round ancillary gauges in place of the pie-wedge gauges on earlier models. These were situated to the left of the big centrally located tachometer. The seats were re-contoured for better lateral support.

To reduce noise, vibration, and harshness, Mazda put ball joint mountings on the RX-7's anti-roll bars. Vehicle-speed-sensitive power steering was used on all models but the GTU, which had engine-speed-sensitive power assist.

The new GTU^S was almost, but not quite, the Turbo without the turbo. The GTU^S shared suspension and brakes with the Turbo, but not that model's standard ABS. Like the Turbo and convertible, the GTU^S had an aluminum hood for a lower overall weight. New "multi-x" aluminum alloy wheels, lighter than previous Turbo-model wheels, measured 7.0x16in and wore 205/55VR-16 Bridgestone Potenza RE71 rubber. While the GTU^S had a 4.30:1 final drive ratio, the Turbo shared the 4.10:1 ratio of the other manual-shift RX-7s (automatic-equipped models came with a 3.909:1 final drive, with the exception of the convertible, which was 4.10:1). The Turbo, however, had its own gearsets, while the GTU^S shared ratios—except for that of fifth gear—with the other manual-shift RX-7s.

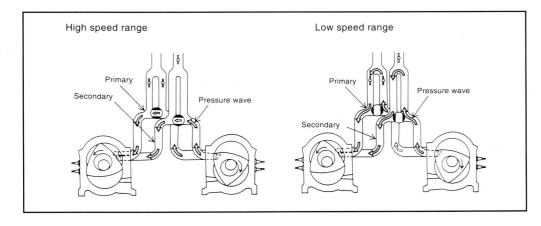

On 1989 and later naturally aspirated engines, the rotary valve allowed a shorter route for dynamic effect ramcharging at high rpm and a longer route at lower rpm. Mazda

The Turbo and the GXL had a new rear parcel shelf that hid items in the "trunk," and lifted out of the way when the rear hatch was raised. A six-speaker AM/FM stereo system was standard on all models, and the GTU, GXL, Turbo, and convertible added an auto-reverse full-logic cassette player. A compact disc player was optional only on the Turbo and convertible. All 1989 RX-7s got a larger 18.5gal fuel tank and, except for the convertible, "mouse-type" automatic shoulder belts.

The improvements to the engine made for faster-accelerating and more fuel-efficient RX-7s. EPA city ratings were 17mpg for all except the Turbo, which was rated at 16mpg. Naturally aspirated RX-7s with a five-speed returned 25mpg on the highway test, and the automatics and the Turbo hit 24mpg.

Acceleration data varied. Mazda, for example, claimed 0–60mph for nonturbos at 7.7sec, while a GTU^S tested by *Road & Track* turned an 8.6sec run. Likewise, a quarter-mile time of 15.9sec was claimed by Mazda, while the magazine recorded 16.7sec. *Road & Track* and Mazda both agreed on a 6.3sec 0-60mph time for the

Turbo, while the magazine's quarter-mile of 15.1sec (at 93.5mph) compares well with Mazda's 14.9sec claim. The convertible paid for its style, however, with a 11.3sec 0-60mph sprint and a relatively leisurely 18.4sec quarter-mile.

If turbocharging wasn't enough, however, Mazda was working with a production-prototype three-rotor engine, even bringing an RX-7 prototype to the U.S. in late-1988 for testing and giving a few journalists an exciting ride. While multirotor Mazda racing engines and prototypes by Mazda, Mercedes, and General Motors had been built, no multirotor Wankel engines had ever seen serial production. Although it would seem to be simple enough to just stack on rotors—often cited early on as an advantage of the rotary—it's not so easy in practice.

The problem is the eccentric shaft, the rotary equivalent to a crankshaft. It's easy to slide a rotor onto either end of a single-piece eccentric shaft, but there is no way to get a rotor onto the center throw of a three-rotor shaft. For the three-rotor, Mazda used a built-up eccentric shaft, which had an extended "nose" with a taper and a keyway. An end piece was fitted to the extension. The third rotor thus transmitted power through the nose of the eccentric and the extension, a more sturdy arrangement than a joint between the rotors. The eccentric, of course, has its throws at 120deg instead of the 180deg of the two-rotor. Mazda's four-rotor race engines were built up, not by joining two two-rotor engines, but by using the three-rotor pattern on both ends of a two-rotor.

The three-rotor was installed in an otherwise stock left-hand-drive Turbo chassis, but without the intercooler hood scoop. The result, wrote Don Fuller in *Motor Trend*, was an engine with the smoothness, power, and torque of a V-12. It would take full throttle at 1000rpm without sputtering, and with 220hp at 6500rpm, 14.2sec was the projected quarter-mile time, with a top speed of 155mph. "World-class git-go," said Fuller. Most intriguing was the confession by

This 1991 RX-7 Turbo wears a factory air dam. Note the embossed "Mazda" that replaced the decal beginning in 1989.

"Mazda people" that the "three-rotor would probably appear in a variation of the current RX-7 in 'less than two years.'"

Indeed Mazda was hot, so hot that the three-rotor prediction was met with nothing but anticipation. In July 1989, Mazda put the MX-5 Miata on sale to rave reviews from press and public. Sales for the cute little roadster were forecast at an annual rate of 40,000, and it proved a fairly accurate estimate. George McCabe, then vice-president of sales and marketing for Mazda Motor of America, predicted RX-7 sales would be 20,000 per year. But while sales for MMA in 1989 totaled 263,378, the RX-7 sales tapered to 16,249.

1990

George McCabe had told journalists that the introduction of the Miata was the beginning of the "New Mazda," a "turning point for Mazda in this country." Indeed there were many new Mazda products on the way, with the new 323/Protege debuting and a new 626 and 929 in the wings.

But there was little to talk about in the way of new product or features for the RX-7. Models were limited to the GTU, GXL in two-seater or 2+2 versions, convertible, or Turbo, with a choice of five-speed manual or four-speed electronically controlled automatic transmission for all models but the Turbo.

One anomaly, however, was the GTU^S. The decision was made to drop it from the 1990 line but not before exactly one hundred had rolled off the production line. Thus, no advertising mentioned the model, nor was it cited in the material given to the press. The only public reference to the 1990 GTU^S was a footnote in the specifications listing in the 1990 RX-7 owners manual, citing the GTU^S axle ratio.

For the models Mazda acknowledged, one technical change was the switch from vehicle-speed- to engine-speed-sensitive power steering on the GXL and convertible, a feature previously used only on the GTU. Only the Turbo was left with what Mazda described as "sophisticated vehicle-speed-sensitive power assist," and Mazda gave no explanation for the change. A change in accessories was making the full-logic auto-reverse cassette player standard across the board and the compact disc player standard in the convertible. The convertible also got a driver's side airbag.

Despite being the most refined and developed RX-7 to date, sales in 1990

The base model for 1991 was called simply the RX-7 Coupe.

plummeted to 9,743. Undoubtedly the Miata pirated sales from the RX-7, and the ever escalating price of the rotary sportster did not help either. Mazda was anything but worried, however, at least in its public pronouncements. It credited the Protege, 626, MPV, and the Miata with strong showings, as well as the new Navajo sport utility, which was named *Motor Trend*'s "Truck of the Year." Other Mazda models won accolades as well, even the venerable RX-7. Said Clark Vitulli, chief operating officer of MMA, "Given the gloom and doom atmosphere surrounding the auto business, this passenger car sales record and our strong showing in truck sales is a tribute to the quality of Mazda's product line and the abilities of our dealer organi-

Mazda revised the instrument panel for the RX-7's final year with a 160mph speedometer, 9000rpm tachometer, and subsidiary gauges pointing to eight o'clock at rest.

Infini IV: The Orient Express

The first Infini was a 1987 Mazda 323. But it was more than an ordinary 323, more than the turbocharged 323 GT already on the market in Japan, even more than the all-wheel-drive GTX pocket rocket. The Infini was a special version of the latter. "Tenacious" was the adjective of choice for this Japanese family sedan turned Bad Boy.

Flash forward from that time to 1991. Infini had become synonymous with Mazda "skunk works" projects, and a special RX-7,

No badging on the Infini IV identified it as an RX-7.

the Infini IV, is the ultimate production Infini and the ultimate RX-7, at least in terms of performance. Essentially it is the Turbo with more power and less weight. As a more elemental RX-7, Mazda said it offered a glimpse of the future RX-7, due for a 1992 introduction.

Eager to inform (and get some ink in the enthusiast magazines), Mazda shipped over one Infini IV for a limited stateside stay. So limited, in fact, that only *Road & Track* and *AutoWeek* got a look at the car before it was shipped back home. Whether this short visit helped to reveal the third generation is debatable, but it certainly frustrated Americans who wanted just that much more from the current-generation RX-7.

The Infini IV ("RX-7" actually appears nowhere on the car) was a true Japanese-market car, and as such had right-hand drive. Once acclimated to this "inverted" driving position, American testers most noted that this was indeed an RX-7 with a sharpened edge. The cockpit was tighter, with bolstered, hard-to-get-into but snug-as-a-baseball-in-a-glove seats. The seats were not adjustable for rake, limiting headroom for taller drivers, but were lighter than the cushy stock seats.

The compact cockpit was exaggerated by rectangular pads on the center console and door which served as braces for the driver in hard cornering. A foot rail was added on the left side for passenger comfort. A final touch was a fat-rimmed Momo steering wheel. And for the dedicated sports car types, the otherwise standard air conditioning was deleted. Every ounce must go!

Mazda engineers dropped weight from the Infini by removing insulation, which quiets the beast but adds nothing to performance. A lighter, less restrictive and noisier exhaust system helped bring weight down to 2,700lb, which was 287lb less than the U.S. specification RX-7 Turbo. The exhaust also let the turbo pump the rotary for another 15 horses over the American market 200hp rating. The drivetrain, with a five-speed manual transmission, was otherwise stock except for the Torsen limited-slip differential.

Suspension modifications included spring rates increased 10 percent and shock absorbers 25 percent stiffer on compression and rebound. Less compliant bushings were also used, including rear suspension bushings that defeated the DTSS passive rear steering suspension. A strut bar (a brace between the front strut towers) stiffened the chassis. With the stiffer springs, a smaller, 23mm, anti-roll was fitted up front. Overall, the suspension sets the Infini IV about 1in lower than standard ride height of U.S. models.

Tires developed by Pirelli specifically for the Infini IV, P-Zero 205/55-16s, were mounted on standard-pattern BBS wheels.

Harder brake pads were installed in the four-piston front and single-piston rear disc brakes borrowed from the Turbo. Instead of the speed-sensing electronically controlled power steering of the Turbo and other upgrade RX-7s, the Infini had the engine-speed dependent power steering from the base model. And although the ratio was slowed from 15.2:1 to 17.4:1, the simpler system, according to Mazda engineers, gave better feel for autocross and racetrack events.

Externally the Infini IV was identified by an infinity symbol decal above the left taillight, infinity symbol hubs on the BBS wheels, amber fog lamps, and obligatory green-black paint. It was priced about $2,500 above an RX-7 Turbo in Japan, the only place it was sold. It wasn't a homologation special, since there was no appropriate racing class in Japan for the model. It was just, well, a hot rod.

Test drivers noted the firmer suspension and lack of sound deadening. The tur-

The Mazda Infini IV was touted as a harbinger of the RX-7 to come.

bocharger's whistle and the rotary exhaust note stood in for the missing radio. *Road & Track* liked the sticky Pirellis and prodigious cornering ability. Testers described the car as "neutral as Switzerland," and it posted a 0.91g on the skidpad. It took 7.0sec to go 0–60mph, and the quarter-mile required a mere 14.9sec at 94mph. Stopping was impressive, too, as the discs hauled down the two-seater from 60mph in just 141ft.

"Through the years, we've seen sports cars grow larger, softer, and heavier," said *Road & Track*, "which makes Mazda's reversal of the trend even more exciting. Let's hope that this Japanese-market Infini's chassis is truly representative of the new generation's, and that Mazda's stylists have created a shape worthy of the mechanicals."

Let's hope that *Road & Track* had a firm grip on its collective hat, because if the magazine's staff liked the Infini IV, they were going to love the new RX-7.

The Infini IV's interior was crafted for the performance driver, although as a Japanese market car, the steering wheel was on the right.

The Infini IV's engine received a slight power boost, and the strut tower bar increased chassis stiffness.

zation."

1991

Entering its sixth model year and with an almost entirely new version of the RX-7 just over the horizon, there was little new to expect from what would be the model's last year. Yet refinements and the addition of across-the-board standard equipment would make the 1991 RX-7 possibly the best yet.

The GTU and GXL were merged, losing the fender decal and becoming simply the RX-7 Coupe with lots of standard equipment and new five-spoke aluminum alloy wheels. Every 1991 RX-7 would come with air conditioning and power windows and door locks, and all steering wheels and shift knobs were leather-wrapped.

Mazda also grouped options into "packages." On the Coupe, Package A included upgraded interior trim, a power sunroof, tilt steering, adjustable lumbar and thigh support on the driver's seat, cruise control, and four-piston front brake calipers and ventilated rear discs. Package B, also available on the Turbo, included leather seat trim, a compact disc player, and a rear cargo cover (standard on the Turbo). The fully equipped convertible had a four-speed automatic transmission as its sole option. Even the compact disc player (with headrest speakers) was standard.

Also standard on the convertible was new Hi-Reflex paint. This process rotated the body while it was baked in the drying oven, allowing the use of thicker paint without runs or drips and resulting in a smoother, glossier finish.

Despite the improvements, RX-7 sales fell to 6,986 for calendar year 1991, with Mazda's total automobile sales trailing off slightly as well. Had Vitulli spoken too soon?

1992

There was no 1992 RX-7. Mazda dealers simply sold the 1991 RX-7 until there were no more. The all-new 1993 RX-7 debuted at the Los Angeles Auto Show on January 2, 1992, and went on sale in the spring. Sales totals for the RX-7 were down for the year, despite the new model, totaling 6,006 for old and new models combined. It was an ominous sign, considering that a new model usually means a surge in sales.

Chapter 13

RX-7 Competition 1989–1993
Less is More

Price Cobb in the four-rotor RX-7 GTO racer.
AutoWeek/Richard Dole

With the changes occurring at Mazda, racers would experience repercussions not so much from the racetrack as from the Boardroom. Mazda had reorganized its corporate structure in the U.S. by consolidating Mazda East and Mazda Central with Mazda Motor of America, Inc., headquartered in Irvine, California. Beginning in July 1988, a corporate restructure put racing under sales promotion which was overseen by Dick St. Yves. It was a new field for St. Yves, who was a sports enthusiast but who previously had not been directly involved with racing.

St. Yves would change the *modus operandi* of Mazda's competition department. Instead of providing "seed money" for a variety of projects, Mazda would support intensive competition efforts that would provide maximum exposure for Mazda products. It would mean the end of Jim Downing's Camel Lights program,

> *"The driver asked for more than he should have. He pressed the right pedal farther than the tires wanted him to."*
> —racer Jim Mederer commenting on his 240mph Bonneville crash

which even at best was a spear carrier for the GTP prima donnas. To encourage sales, the MX-6 would be featured in GTU, while a new four-rotor RX-7 GTO program would replace the horsepower-shy three-rotor effort.

Only four Mazdas were entered in GTU in the 1989 Daytona opener, but it looked like Amos Johnson, running an RX-7 since his new GTU MX-6 wasn't ready, was about to score a fifth consecutive Daytona 24-Hour win. Leading with but four hours to go, ignition problems required a twenty-one-minute fix. Passed in the pits, Johnson—with co-drivers Dennis Shaw and Bob Lazier—finished third.

Fortunately, Al Bacon, with co-drivers Bob Reed and Rod Millen, saved Mazda's, well, bacon. Bacon, slated for another GTU MX-6, had put his RX-7 on the GTU pole. By winning the class he marked Mazda's eighth consecutive GTU victory at Daytona's round-the-clock race.

The only other 1989 GTU win in an RX-7 was John Finger's Miami reprise, the mid-season MX-6 arrival yielding another

Peter Farrell and Willy Lewis drove third-generation RX-7s to fifth- and sixth-place standings in the Bridgestone Potenza Supercar series. James Resnick

Mazda GTU championship. Mazda recorded more firsts than any other make, but Bob Leitzinger took the driver's title in his Nissan.

Roger Mandeville persevered in the three-rotor RX-7 in GTO, but a fifth at Daytona (with Kelly Marsh and Brian Redman) and three other fifth-place finishes were the best that could be tallied before the plug was pulled in mid-season.

In amateur racing, the SCCA's runoffs saw an incredible slugfest in GT-2 between the Porsche 944S of David Finch and the RX-7 of John Finger, at least for the first twelve laps. Finger went off-roading to pass but broke his RX-7's suspension, giving Finch the win. Matt Mnich brought his RX-7 from last to second after a first-lap tangle put him off the road.

In GT-3, Bill Reid had the only RX-7 in the race, but his second-fastest qualifying time turned into a DNF from an off-course excursion.

1990

Cinderella stories are intriguing because they don't happen very often. Rags to riches is rare. So are debut wins for radically new race cars. And it's rarer still for a novice race car to win a 24-Hour race. Still, when a new four-rotor RX-7 qualified on the pole for the 1990 Daytona opener and then set the fastest race lap, a victory seemed very possible. Success loomed large for Pete Halsmer's GTO RX-7 #1—one of the two four-rotor cars—until it ingested a small bolt from the air cleaner, which destroyed the engine and eliminated the car from the race. So that Halsmer could get points toward the driver's championship, he was given seat time in the #63 car, for which the scheduled team had been Jim Downing, Amos Johnson, and John O'Steen. The #63 car finished seventh overall and, more importantly, second in GTO.

The two four-rotor RX-7s represented a change in philosophy for Mazda. Realizing that small independent teams couldn't defeat the likes of the Roush-prepared, Ford Motor Company-backed Mustangs, Mazda Motor Company in Japan funded the cars directly. The new plan was masterminded by competition manager Ken Kinoshita. Pete Halsmer, the 1989 GTO champion in a V-8 Cougar, was hired as lead driver.

With even amateur race cars being tube-framed replicas of their production equivalents, it's no surprise that the IMSA GTO was anything but a front-engined GTP car in an RX-7-shaped body. Master race car designer Lee Dykstra sketched the RX-7's chassis, a steel tube frame with carbon fiber and aluminum panels. Like other cars in the class, the roof and windshield were the only actual production parts on the car. The rest of the body "represented" the RX-7 body but, even as a fiberglass replica, was significantly different.

Suspension was pure racer, with double A-arms front and rear and inboard pushrod-actuated coil-over shocks up front. Coil-over shocks were high-mounted at the rear. The ventilated discs had six-piston calipers in front. Sixteen-inch wheels were used at first but were later changed to seventeen-inchers. A new five-speed Hewland transmission was front-mounted, and the four-rotor's power dictated a triple-plate Borg and Beck clutch.

The four-rotor engine was designated the 13J-M2, the same engine that had powered a Mazda 767B to seventh over-

Peter Uria bought Amos Johnson's Team Highball GTU RX-7 and used it for a 1990 Daytona 24-Hour win. AutoWeek/Richard Dole

Amos Johnson, with Shaw and Lazier codriving, lead Daytona until the twentieth hour when ignition problems knocked them back to a third place finish. AutoWeek/Richard Dole

all at Le Mans in 1989. The four-rotor was similar to the three-rotor engines in that it was essentially a two-rotor engine with a single rotor added to each end using the "tapered coupling method." Because the center side-housing is not load bearing (there's no center "main bearing" even on the two-rotor engines), Mazda saved weight by making this piece aluminum.

For greater wear resistance in high rpm and temperature conditions, aluminum and cast-iron side-housings were gun-sprayed with Cermet. Apex seals were made from a high-heat conductivity silicon nitride ceramic, and the trochoic housing's surface was plated in micro channel chrome-molybdenum instead of the solid plating used on earlier race engines.

The 13J-M2 had special electronic fuel injection with a unique two-step variable induction system. A sleeve in each of the four intake tracts opened to shorten the tract length at higher rpm. Set to open at 7500rpm at 80 percent load, the shorter tract length tuned the intake resonance for higher output at the higher engine speeds. The 13J-2M rotary was rated at over 630bhp at 9000rpm and more than 378lb-ft of torque at 8000rpm; all this from 2616cc and on regular unleaded gasoline.

As usual, after Daytona came Sebring, and at that enduro it was car #63 that expired while #1, with Halsmer, John Morton, and Elliot Forbes-Robinson, stayed out front until a crash near the end put the car out. Their lead was sufficient, however, for a third-place finish. Seconds at Miami and Long Beach were

Wearing #1 from his 1989 championship season, Pete Halsmer debuted the four-rotor RX-7 at Daytona in 1990, but never drove the car in the race. AutoWeek/Richard Dole

After finishing second at the SCCA's runoffs in 1991, Michael Lewis won GT3 in 1992. AutoWeek/Richard Dole

followed by the four-rotor's first win at Topeka. That and another win at Mid-Ohio were followed by the San Antonio street race where Halsmer and the four-rotor passed an RX-7 milestone, the 100th IMSA win. However, the 1990 GTO championship would fall to Dorsey Schroeder, and Lincoln-Mercury got the manufacturer's title. Halsmer placed a close third in the standings.

Roger Mandeville managed the factory MX-6 GTU team, but it was Amos Johnson's old Team Highball RX-7 that won GTU at the Daytona 24-Hour. It had been leased to Peter Uria, who had Bob Dotson, Jim Pace, and Rusty Scott as co-drivers. But the Daytona 24-Hour was the last RX-7 win of the year, while the factory MX-6 took GTU manufacturer's laurels and first and second in driver's standings.

In the SCCA's amateur competition, John Finger's RX-7 was reportedly assembled from five cars, "with only Finger's experience holding it all together." Finger also held second place until a full-course yellow went green with two laps remaining. The resultant free-for-all saw Matt Mnich lead to the finish with Finger coming home third.

1991

Mazda had but a sole GTU win in 1991, having shifted all attention to the GTO four-rotor effort. But the GTU win came, surprise, at Daytona, where Dick Greer and teammates brought in the ex-1989 Daytona-winning RX-7 one hundred miles ahead of the second-place GTU car. Greer's co-drivers were Al Bacon, who had sold the car to Greer, and Mike Mees and Peter Uria. The second-place GTU car was an ex-factory MX-6, and Roger Mandeville, along with teammates Amos Johnson and Kelley March, finished third in a new GTU RX-7. Third would be the highest any RX-7, or any other Mazda, would place in GTU for the rest of the season. And third would be Mazda's final standing in the GTU manufacturers championship, well behind a Nissan-Dodge donnybrook.

There was little to gain from another GTU championship, however, and much from a title in GTO. But the technology required to be competitive in the GTO class meant that only the factory teams from Ford and Nissan—and Mazda if willing to play—had any chance of winning.

Mazda came back cocked and loaded from the between-season break. The car had undergone continual development during the 1990 season, and more changes were made during the short off-season. The suspension was extensively modified, and the body shape changed for greater downforce. About the only thing not changed, Dykstra later said, was the frame.

A two-car team was entered again, with Pete Halsmer and Price Cobb as the regular drivers. In the season opener at Daytona, however, Cobb and Halsmer wound up sharing the same car after Halsmer's car was crashed by a teammate. The pair, along with Brian Redman and John O'Steen, struggled to bring the car home for a fifth-place finish.

Mazda, however, claimed an early lead in the points standings, as both Halsmer and Cobb tallied wins. Although the RX-7s had begun at 2,250lbs and IMSA had added weight to keep the racing competitive, yet another 100lb was added when Cobb won the New Orleans street race. Although Cobb managed one more win—at Laguna Seca where he avoided an oil slick which Jeremy Dale and Pete Halsmer didn't—Mazda had no other firsts in GTO after the weight was added.

Mazda still had the lead in manufacturer's points, however, and Halsmer had the most points toward the driver's championship as the season wound down to the finale at Del Mar. For Mazda to win the manufacturer's title, all Cobb had to do was finish ahead of the Nissans, and Halsmer could win his second GTO championship by finishing in the points. They both did. Cobb finished second, behind Gordon's Mustang but ahead of the two Nissans, and Halsmer stayed out of trouble to finish sixth. It was good enough: Mazda and Halsmer had championships in IMSA GTO. However, it was also the end of the four-rotor GTO program. Mazda retired the cars, one going directly to Japan for museum display. There would be no reprise in 1992 as Mazda had a new project: IMSA GTP.

At the SCCA's runoffs it was a year for contenders but not champions among those driving RX-7s. In GT-3 Michael Lewis finished second, while Matt Mnich, who had not been in striking distance for most of the race, took home the third-place medal in GT-2.

And for proof that good ideas are never out of style, there was the outright win at Pikes Peak in the open class for Rod Millen's 4WD RX-7 turbo.

1992

Mazda won Le Mans in 1991, but because the French effectively banned rotary engines from future competition,

Dick Greer placed first in GTU at the Daytona 24-Hour in 1992 and 1993, adding Sebring in the second year for good measure. AutoWeek/Richard Dole

Mazda stepped into GTP, IMSA's center ring, with an adaptation of the Le Mans racer.

The RX-792P never really came together, however, as it was plagued with problems and down on power. Nor was the factory team ever really competitive in the Bridgestone Potenza Supercar series. This "showroom stock" class for exotic cars featured the use of its sponsor's tires but the rules called for "equalizing" the competing marques by ballasting the faster cars. The two-RX-7 team featured Peter Farrell and Willy Lewis as drivers, and for their first year in the series, they finished fifth and sixth overall.

IMSA GTU would be a Nissan show, but the Mazda stranglehold on the Daytona 24-Hour opener was as strong as ever. Dick Greer, again with the team of Bacon, Uria, and Mees, won the 'round-the-clock affair. Actually "won" is something of an understatement: The second-place GTU finishers, the Alberti Motorsports/Team Peru MX-6 of Eduardo Dibos, Juan Dibos, and Raul Orlandini, was sixty-three laps behind.

At the SCCA's runoffs at Atlanta, Bill Reid's GT-2 RX-7 wound up in a battle for third with a Porsche 914 and a Sunbeam Tiger (talk about variety) that had everyone on their toes to see who would emerge first from under the bridge on the last lap. It was Reid, holding on to win the bronze. In GT-3, Michael Lewis hounded the Corolla of Pete Peterson until the last lap when he got his RX-7 around for the win. Stacy Wilson finished third in another RX-7.

In the SCCA's World Challenge, Makato Yamamuro finished second driving an RX-7 Turbo in Class B, and Sergio Afansenko in an RX-7 was fifth in Class C.

Disaster struck at Bonneville when Racing Beat attempted an encore of its earlier record runs. This time a third-generation RX-7 would be used, competing in C-BMS, or blown modified sports. Class rules allow a modified chassis and an all-out race engine. The racer was as sophisticated as anything Racing Beat had ever made. Its three-rotor 13G engine was boosted by three high-volume KKK turbos, each with an intake port of its own and a water-to-air intercooler filled with ice. Output was 774hp at 8100rpm. The goal was 300mph. Racing Beat boss Jim Mederer wanted to be the

Nice guy Pete Halsmer finished fifth at the 1991 Daytona opener but would go on to win the GTO championship for himself and for Mazda the manufacturer's title. AutoWeek/Richard Dole

first to break that barrier in a "doorslammer," as the lowered but stock-bodied cars are called.

Bonneville Speed Week is only one week long, shorter if the weather's bad. So for experience with the car at speed, Mederer, who would drive at Bonneville, took the car to the high-speed track at the Transportation Research Center in Ohio. It was a handful at 215mph, with too little aerodynamic downforce at the rear. A second test, this at the El Mirage dry lake event, shorter than Bonneville, yielded only 182mph. The RX-7, even with 150lb of ballast at the rear, was traction limited.

Then came Bonneville. Aerodynamic adjustments trimming allowed 225mph on Mederer's first run. But there were two days of rain delay—mostly waiting for the course to dry—before the RX-7 would be allowed on the long course, where records were set. And then, during the RX-7's first run on that course, at between 230 and 240mph, "it became an airplane," as Mederer described it. Turning sideways, the car soared ten feet into the air, came down on a rear corner and then, before finally stopping upright on its wheels, did more flips and rolls than high-diver Greg Louganis could imagine. The car, except for the roll cage, was destroyed. Fortunately, Mederer was uninjured but took the blame, telling *AutoWeek*'s J.P. Vettraino that "The driver asked for more than he should have. He pressed the right pedal farther than the tires wanted him to."

1993

Slow sales put an end to fast cars, at least for the duration. Mazda reluctantly closed down its racing program for 1993, suspending the two programs in which it participated actively—IMSA's GTP and the Bridgestone Supercar series—but saving the contingency program for privateers driving Mazda products.

Mazda was still a force in GTU, however. And like deja vu all over again, the Greer RX-7 won the Daytona 24-Hour and, for good measure, the Sebring 12-Hour event. Dick Greer even repeated the crew of Bacon, Uria, and Mees for Daytona. Despite Nissan-mounted Butch Leitzinger's six wins, Greer finished a close third in the 1993 driver's championship, and the next four places in the standings went to Mazda drivers as well.

Joe Danaher and Peter Farrell entered Mazda RX-7 Turbos in IMSA's Firestone Firehawk spec tire series and finished a close second and third, respectively, to Peter Schwartzott's Honda Prelude VTEC. David Lapham meanwhile ran a competitive campaign with an RX-7 Turbo in SCCA's Class B World Challenge series but ended up seventh. In Class C, Gary Ain drove his RX-7 to a sixth place in the standings.

And finally, at the SCCA's runoffs at Atlanta, it was a disappointing year for Friends of the RX-7. Although Bill Reid and Mike Green took the top two grid spaces in GT-2, Reid was black-flagged for pacing the field too fast, finally finishing fifth, and Green struggled to sixth after choosing a too-soft tire compound. The best RX-7 finish was the fourth place of Michael Lewis who, running the same car in GT-2 and GT-3, ran into more problems than time to solve them. The GT-2 placing was better than GT-3 where the defending champion slid off the track in lap one, turn one. One had to go to twelfth in the GT-3 standings to find the highest placed RX-7, that of Cameron Worth.

Chapter 14

1993–1994 RX-7
Leaner and Meaner

Uncompromised. Think about how good that sounds. Purity of purpose, clarity of objective, designed to be very good at one thing: A one-trick pony. Yes, but as Paul Simon sings, he does it very well.

That was the 1993 RX-7. It was oh so good. And it came from being uncompromised. Few other machines are so purely purposeful. Atomic submarines, maybe. An F-16. A Patriot missile.

Credit project manager Takaharu Kobayakawa on this one. And Kenichi Yamamoto, president and chairman of the board. And a belief in the concept and an eye on the prize.

But what was the concept?

Takaharu Kobayakawa took over the RX-7 program from Akio Uchiyama, who had led first and second generation de-

> *"The skin stretches tightly over big tires mounted on wide alloy wheels stuck way out at the edges. There's a sense of the muscular minimalist purposefulness of early Ferraris and Cobras."*
> —Motor Trend on the 1993 RX-7

velopment, in January 1986. It would be up to "Koby" to continue development of the second generation RX-7, including introduction of the convertible, and the "minor change" scheduled for the 1989 model year, and the development of the successor to the second-generation RX-7.

Kobayakawa established a development team for the new sports car, beginning a "deep probe" into just what that sports car could be. Engineering should be such fun: visits to museums in the West, from fine art to automotive and aeronautical, driving lessons at Jim Russell's driving school (including taking a Russell Mazda-powered formula car back home), and compiling a comprehensive collection of sports cars, encompassing even a few Italian and German exotics, as well as a fleet of high-performance go-karts. More fun indeed than slide rules and computer printouts.

Yet much was riding on the project. The RX-7, if not the sales leader and profit

The stunning new RX-7 appeared as a 1993 model. Pop-up headlamps were matched by fog lamps in the radiator opening.

Put me in coach, I'm ready to play.

generator the 323 and 626 had been, was a very high profile automobile for Mazda. Kobayakawa, an engineer, was an early member of Yamamoto's rotary engine research department (from April 1963) and, serving in Mazda's California technical liaison office, had been in the United States during the oil crisis debacle in the mid-1970s. He had, as director of Toyo Kogyo's public relations, launched the 323/GLC, the 626, and the original RX-7. And now, again back in engineering, he was responsible for the company's flagship sports car.

What indeed would the new sports car be?

As with P747, the team established concept before hardware. Even the sacred rotary engine was not compulsory. The team targeted four prime factors: direction (a "pure" sports car), spirited (with the vitality of an untamed stallion), alluring (tempting, enticing, and seductive), and excitement (arousing the "emotion of motion"). Practicality and Comfort were secondary factors.

Although free to choose other engines, Koby's team chose the rotary for the same reason it was used in the original RX-7: it was a compact source of substantial horsepower. And anyway, the rotary was the soul of Mazda, its reason for being. As Yamamoto would note, "To forsake the rotary would be losing our own identity."

Yet there remained a choice of which rotary and its location in the car, whether traditional front mid-engine or true mid-engine. Kobayakawa adamantly refused the multirotor engine, despite pressure from his boss, Yasuo Tatsutomi, head of product planning and development. Tatsutomi, as a rotary engine engineer had helped make the multirotor engine feasible. The multi, in three- and four-rotor versions, had been used successfully in racing and prototype production versions and had been developed and shown to the press in a second-generation RX-7.

Kobayakawa knew that a triple would not only be a heavier engine, but its power output would require a stronger and therefore heavier chassis. A triple would also need more cooling and a bigger, heavier radiator. And as weight was added to the powerplant and accessories, even more weight would be needed for the chassis to support it, the brakes to stop it, and on and on. What Kobayakawa did not want was "Zevolution," that process that had changed the originally relatively Spartan Datsun 240Z into the mid-1980s boulevardier 300ZX. Fortunately, the new Cosmo would provide a home for the production three-rotor, though only for the Japanese market, taking the pressure off Kobayakawa for a three-rotor RX-7.

Engine position was another matter. There had been mid-engined Mazdas before, from 1965's R16A (with a 4x400cc rotary engine) roadster, to the RX500 shown at the 1970 Tokyo Motor Show, to a variety of sports racer competition cars. Yet practicality, particularly the limited luggage room, the bane of virtually every mid-engined car, and tradition argued for the front engine position. Mazda had been successful with that layout, in competition and in the showroom. Why tamper with success?

To be a world-class sports car, Kobayakawa believed that the minimum power to weight ratio was 5kg (11lb) per bhp. The best Mazda had yet achieved, the 1990 Infini IV, was only 5.7kg per bhp. While part of the equation was chassis weight, which Kobayakawa wanted to limit to 2750lbs (1250kg), the other half was horsepower. Kobayakawa figured that to achieve sufficient power the rotary would need to produce a minimum of 250bhp, but no massaging of the current production turbo rotary would produce that kind of output.

The answer was to double up on turbos. Mazda engineers wanted more than mere peak output, however, so they developed a complex computer-controlled sequential deployment of the turbos. Some of the development was shared with the Cosmo, which would be available with either a twin-turbo two-rotor 13B or a twin-turbo three-rotor 20B.

But unlike the Cosmo, which used two different turbos sequentially, the RX-7's 13B-REW (REW for, curiously, Rotary Engine Double-turbocharger) used two identical Hitachi HT12 turbos with 51mm, 9-blade turbines and 57mm, 10-blade compressors. The Cosmo's primary turbo had flat turbine blades to maximize the impact effect of the rotary's exhaust pulse, but the RX-7's primary turbo, like the secondary on both cars, had curved "high-flow" blades for rapid spooling at high rpm. The turbos were also mounted on a new exhaust manifold that placed them even closer to the exhaust port, raising low-speed boost up to 25 percent.

The sequential operation of the twin turbochargers was more complex on the RX-7 than on the Cosmo. At low speeds, most exhaust was routed to the primary turbo which spooled rapidly to provide

In late-September 1987, finalists were chosen from scale models and full-size side drawings, with the California entry (center red) a leading contender. Mazda design general manager Shigenori Fukuda (right) discusses entries with Yoichi Sato (in jacket) and Tom Matano (black shirt). Mazda

A computer-controlled machine carved a full-size model for wind tunnel testing; the first full-size clay model differed significantly in detail from the final product. Mazda

maximum boost. A small portion of exhaust pressure, however, was bled to the secondary turbo's turbine to keep it spinning in a pre-operation mode of about 100,000rpm. The secondary turbo's compressor rotated as well, but its output was diverted by a by-pass valve to a closed loop. To transition smoothly to full boost from both turbochargers, the secondary turbo was put into a "surge" condition by closing the by-pass valve and spinning the compressor in a closed chamber. Although prolonged surge would damage the compressor, used momentarily it spooled the turbo to 140,000 rpm. Then full exhaust was routed to the turbo and the compressor outlet opened to allow the secondary turbo to provide full boost. Nominal compression was 9.0:1; a wastegate limited maximum boost to 22.4in of mercury (9.6psi).

Depending on the model, the intake charge was directed from the turbos to either one or two air-to-air intercoolers, located in the left or both front corners. Heated air from the intercoolers exhausted through vents behind the front wheels. A two-stage intake manifold, as on the earlier 13B Turbo, was used though with runner lengths retuned. A single primary port for each chamber was located on the central side housing, the end housings containing the single secondary intake port. The exhaust port was, of course, peripheral.

From the turbochargers, the exhaust passed through a single-tube exhaust pipe to a two-stage, five-bed, oxidizing, 3-way, low-restriction catalytic converter.

Ignition was electronically controlled, using a pair of platinum-tipped spark plugs per chamber. The ignition microprocessor retarded the spark on an automatic transmission-equipped RX-7 on signal from the transmission's microprocessor, smoothing shifts by momentarily reducing power.

Maximum engine cooling with minimum drag was achieved by tilting a compact high-efficiency radiator forward 58 degrees from vertical. Two fully shrouded electric fans ran in three stages as needed. Oil coolers were located at the front corners, either both sides or just on the left, depending on model.

The result of all these efforts was 255hp at 6500rpm with 217lb-ft of torque at 5000rpm. The relatively high torque peak shouldn't suggest, however, that the engine was peaky with a narrow power band. It meant that it came on strong down low, particularly for a rotary, and then went on a power trip on top when the second turbo kicked in. *Motor Trend* called throttle response "nearly instantaneous," while *Road & Track* quoted race driver Danny Sullivan as saying it was "excellent for a Turbo."

A real test of torque was the top gear 30–50mph and 50–70mph acceleration runs performed by *Car and Driver*. In a comparison with the 1992 Chevrolet Corvette LT1 and Nissan 300ZX Turbo, the RX-7 came in third with 10.1 seconds, 10.3 seconds, and 10.7 seconds respectively for the lower speed sprint (if lugging these sportsters in sixth or fifth gear at thirty can be called a sprint!). But in the 50–70mph dash, the RX-7 was the clear winner, at 7.5 seconds compared to 8.3 seconds for the Nissan and 10.1 seconds for the Corvette. Of course, no one in the real world would submit any of these cars to that kind of driving, but it shows how well Mazda's rotary engineers did their homework.

As slick as the twin turbo arrangement was, however, it could still be caught out sometimes. *Car and Driver* called the "flow of power... a bit jerky when the throttle is first opened or closed."

The other half of Kobayakawa's performance goal was, of course, weight reduction. Weight was important not only for acceleration but braking and cornering as well. The body shell was entirely new, using a concept Mazda chassis engineers called "Space-Monocoque," a reinforced unit-body with a partially skeletal construction resembling a tubular space frame. Mazda engineers increased torsional and bending stiffness by 30 percent over the preceding RX-7 without increasing the weight of the body.

The RX-7's chassis diet started in late-1988 and was called Operation Z. It has been suggested that the Z stood for the "zenith" of light weight, though it might just as well be a veiled reference to the

The RX-7 in testing at Mazda's Miyoshi wind tunnel. Mazda

RX-7's overweight archrival from Nissan. Operation Z removed as much metal as possible from the body and frame while design engineer Takao Kijima added braces where extra strength was needed. A six-stage process, Operation Z was scheduled to last through late-1990. In September 1990, final testing took place in Europe and in the United States using a lightly disguised prototype called S1.

Unfortunately, on the road—particularly California's San Bernadino Freeway—S1 exhibited "darty" handling that Kijima determined was rooted in chassis flex. This could be corrected by reinstalling a reinforcement at the back of the rocker panel and by further reinforcing the rocker panel with three small inner bulkheads, dubbed "bamboo joints." Although this plan outraged production planners who envisioned several months delay if the proposed changes were implemented, modification of a prototype proved Kijima's hypothesis. The production engineers were given a drive in the modified car and were so impressed they worked extra hard so no time would be lost.

Other chassis reinforcements included three aluminum braces that traversed the central tunnel, front and rear subframes bolted rigidly to the body for added stiffness and strength, a bar running across the car to support the steering column and aid in side impact resistance, and a bar that connected the rear shock towers side-to-side. On the "performance" model, another shock tower bridge spanned the engine compartment.

Chassis rigidity was also aided by Mazda's innovative Power Plant Frame. Developed for the RX-7 though appearing in production first on the Miata, the PPF rigidly connected the transmission to the differential (which, with independent rear suspension, is fixed in location). Resembling a bridge truss, the PPF was made of high-tensile-strength steel 2.6mm thick with an inner frame made of a 2mm thick laminate of vibration-damping plastic sheet and steel. The net result was a very rigid powertrain that received the engine's torque thrust directly, reducing the need to reinforce the chassis to manage these forces. The PPF also resisted windup better than a conven-

Takaharu "Koby" Kobayakawa (right) was a hands-on and helmet-on product program manager. Mazda

Cold weather testing was performed at Kenbuchi, Hokkaido, in the winter of 1990–91. Mazda

tionally mounted differential, allowing smoother starts and also reducing engine and transmission pitch. Finally, it contributed to passenger safety by helping control crush, keeping the rear axle out of the fuel tank in a severe rear collision.

In spite of earlier efforts in active and passive four wheel steering and variable shock control, Mazda chose a straightforward double A-arm layout. The front upper suspension arm was relatively conventional and in fact shaped like the letter A. It was made from aluminum that had been "squeeze-cast," a technique where the molten metal is kept under pressure while it cools, the pressure blowing out any voids. The resulting part is more dense and has strength comparable to forging.

The lower A-arm, which bears the chassis load via a "coil-over" shock absorber, was actually more L-shaped, or like an A skewed forward. Sliding rubber bushings, (rubber bushings around sliding metal collars that relieve the bushings' fore and aft dislocation while maintaining precise location of the suspension arm) were used on the chassis mount end of the upper arm and forward chassis mount of the lower arm. The lower A-arm's rear mount had a fluid-filled bushing to reduce noise, vibration, and harshness.

The suspension upright/hub carrier was forged steel. The steering tie-rod was located ahead of the A-arm and the steering rack mounted to the front of the front subframe, to which the lower A-arm also attached. The upper A-arm was mounted to a reinforced section of the body shell.

Rear suspension was by double A-arms, and though lacking the elaborate DTSS passive rear steering of its predecessor (which racers disabled when the rules allowed), the new RX-7 had a very sophisticated toe angle control calculated into its rear suspension geometry. And of course Mazda had a ready acronym: Dynamic Geometry Control System. Mazda engineers used computer modeling to evaluate suspension travel, varying the length of the suspension arm, the relative positioning of the arms and the elasticity of the bushings.

The rear suspension was comprised of an upper A-arm with load carrying performed by a coil-over shock absorber mounted to the body shell. The lower arm was actually a lateral link (I-arm) to which a trailing link was attached near the wheel-hub end. Sliding rubber bushings were used at the chassis end of the upper arm, while a plain rubber bushing was used on the lower trailing link and a rubber seated ball-joint bushing used at the chassis end of the lateral link. A ball-joint joined the lateral and trailing links. Squeeze-cast aluminum was used for the upper rear A-arm and rear upright/hub-carrier. The lower links were aluminum forgings.

The extensive use of aluminum in the suspension was an effort to reduce unsprung weight for more responsive suspension travel. Ultra-lightweight wheels were developed for the same purpose. Squeeze-cast aluminum 16x8 wheels were standard on all RX-7 models. Not only were these wheels 1in wider than the widest wheels of the previous generation RX-7, at less than 15.4lb each they were almost 9lb lighter. The space-saver spare was also mounted on a lightweight aluminum wheel. And, like the second-generation RX-7, the jack was aluminum as well.

Squeeze-cast aluminum was also used for the large four-piston front brake calipers. The rear calipers were a single-floating-piston design. Both front and rear discs were a big 11.6in in diameter. The discs were ventilated with every other inner fin split horizontally, increasing cooling area and reducing disc weight. The high-performance R1 model had dedicated ducts routing cooling air to the front brakes. Hot air from the oil cooler(s) routed around the front wheel well to the functional exhaust vent in each fender. An advanced Bosch four-sensor, three-channel anti-lock braking system was made standard on the new RX-7.

The shape of the new RX-7 was so natural that it belies the effort that went into it. Credit Yoichi Sato as the chief designer of the new RX-7. Fresh from Mazda's advanced design studio in Tokyo, Sato was reassigned to Hiroshima in April 1987 to head design activities for the RX-7. It must have been at once exhilarating and intimidating to have front-line responsibility for the look of the company's flagship sports car.

With project chief Kobayakawa, Sato established parameters for design: The hood would be almost four inches lower than the current model, the nose almost eight inches shorter, and overall height would be under 45.3in (1150mm). Japanese influence led to

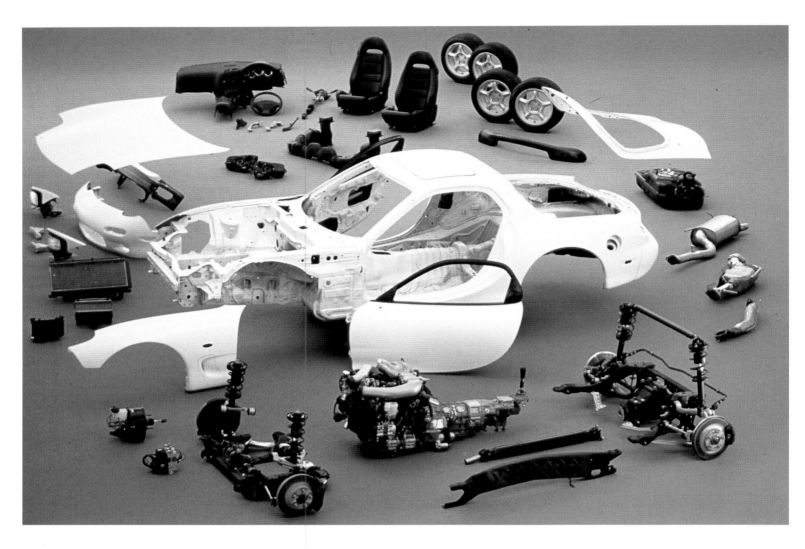

Just your basic RX-7 kit; note power plant frame (black, center foreground). Mazda

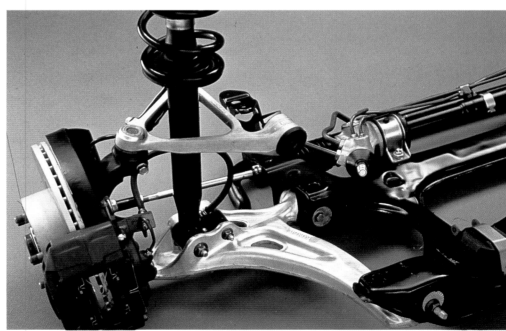

Left front suspension. The lower suspension arm is forged aluminum and the upper is squeeze-cast. Mazda

Rear suspension and drivetrain are secured to a subframe. Mazda

Major components of an "exploded" Mazda 13B-REW engine. Mazda

the project being initiated with a 2+2 version. The third-generation RX-7 was also to be less that 66.9in wide to fit Japan's "small car" category, which would reduce the Japanese owner's annual tax burden considerably, although this consideration was dropped in May 1988 and the RX-7 widened when tax rules were revised for 1989.

The request for concept sketches was met enthusiastically by the four design centers Mazda would employ: Mazda's headquarters design center in Hiroshima, the Yokohama R&D center near Tokyo, Mazda Research of America in Irvine, California, and Mazda R&D in Germany. The last, however, was then no more than a future site, so the independent British design firm International Automotive Design (IAD) was hired to provide a Euro-look RX-7.

From thousands of sketches, eight full size renderings and scale models were assembled in Hiroshima in late September-1987 for the selection of two finalist designs. Chosen were the Hiroshima group's model, reminiscent in profile of the first generation RX-7 and from behind of the Cosmo Sport, and a swoopy design largely the work of Wu-Huang Chin of the Irvine group. Chin's design had a hidden B-pillar, giving the car a "floating roof" look and a short rounded tail, an influence attributed to Chin's own E-type Jaguar.

Both models were refined—Chin's sneaky B-pillar was considered too extreme and the rounded tail aerodynamically unstable—and then both were digitally computer-rendered in full-size clay in Hiroshima. By early-1988 it was time to make a decision. Plastic model shells were made in early-May and both lifelike models sent to Irvine for a meeting of top brass including chairman Moriyuki Wantanabe and president Yamamoto. The difficult decision was in favor of Chin's design, though to be modified with ideas from the Hiroshima version.

Although Chin had established the overall profile and developed the eventual distinctive taillights, missing were the secondary openings in the nose (lacking on the Hiroshima model, too). These appeared on a Phase II model designed at Irvine by the Japanese and American staff working together. A "rotary" shape was also added to the lower lip of the radiator opening.

The completed model was sent to Japan for wind tunnel testing, revised and massaged, and by mid-1989 the design was just about finalized. But the wind tunnel testing hadn't revealed something simulation testing on Mazda's Cray supercomputer did. Turbulence just below the top edge of the rear window was caused by the curve of the roof. It could be eliminated by speeding airflow over this section, and that could be done by lowering the roofline slightly. Not only was this more aerodynamic, but it gave a styling plus as well, making the car look lower with no sacrifice in headroom. And it also hearkened to earlier Mazda design cars and, more importantly to old-time sports car buffs, the "double-bubble" Zagato-bodied coupes of the 1950s. Indeed, double-bubble was the name used for the new roofline internally until someone at Mazda came up with "Aero-Wave Roof"—which enthusiasts would roundly ignore.

It was a smaller, lighter, and tighter RX-7. Weight had been saved by using plastic everywhere from seat bottoms to headlight housings. The retractable headlamps were simpler and therefore lighter than the parallelogram linkage of the second generation's. Even the ignition coil was lighter, reduced from 7.3lbs to 3.3lbs.

And although Mazda claimed it had returned the RX-7 to its roots, such was hardly the case. Surely it was lighter than the second generation and more elemental. The 2+2 version originally envisioned for the third generation evaporated before its debut. But the original RX-7 was a "derived" sports car, descended from the RX-3, from where most of its mechanicals came. That is a sports car tradition. Inexpensive sports cars have commonly come

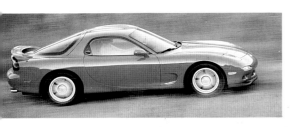

The 1993 R-1 had a front air dam and rear spoiler, unchanged on 1994 R-2 model (shown). Mazda

from sedans. Thus a 1979 RX-7 could be had for about the same amount of money as a moderately optioned Ford Fairmont. But at $30,000-plus the new RX-7 was hardly inexpensive nor was it derived from any existing Mazda. This was no return to basics. This was a whole new ball game, and it put Mazda in the big leagues.

It's U.S. debut was the 1992 Greater Los Angeles Auto Show, where it attracted at least as much attention as the show cars and concept cars, although it had first been seen in the pages of the December 1991 automotive monthlies. *Motor Trend* put a red one on its cover, calling it "NSX KILLER!" *Automobile* said "Rotary rocket science delivers superstar performance." *Road & Track* and *AutoWeek* both sold the "back to basics" line, with the former calling it "Retro Rocket" and the latter, "Mazda administers a dose of the original formula." Only *Car and Driver* prevented a clean sweep of cover photos, going truly back to basics with a Cobra on the cover—but an RX-7 story inside.

These reviews were, as Kevin Smith noted in *Car and Driver*, only on a pair of unfinished prototypes. But still everyone was impressed. Said Smith, "Compared to the current edition, the new car . . . is lighter, lower, shorter, and wider. It looks terrific, lays down a big footprint, seems to handle sweetly, and makes great, edgy shock waves of power: its 255hp rating represents a huge leap from the former maximum of 200hp."

Don Fuller was similarly breathless in *Motor Trend*, "The skin stretches tightly over big tires mounted on wide alloy wheels stuck way out at the edges. The roofline dictates only two people go inside. There's a sense of the muscular minimalist purposefulness of early Ferraris and Cobras, and the pumped-up, rounded fenders and tapering tail define space in the same way as a D-Jag."

In *Road & Track*, Dennis Simanaitis relatively soberly noted, "Its handling felt predictable on rough roads or smooth, its steering exhibiting crispness on input and good communication in response. The ride most definitely emphasized the sporty side of the car's Sports/GT nature. No complaints here." And said Matt DeLorenzo in *AutoWeek*, "The new look works extremely well. Its rounded edges have enough muscle underneath to give the car a strong, purposeful appearance. The RX-7 is compact, neat and clean."

Road tests and intriguing comparison tests were soon to follow. *Motor Trend* offered a match-up consisting of the RX-7 R1, the Mitsubishi 3000GT VR-4, Corvette LT-1, and Nissan 300ZX Turbo. *Motor Trend*'s editor Mac DeMere described the differences among the three versions of RX-7 offered—the base RX-7, the Bose-equipped Touring, and the performance handling R1—by saying, "If, while driving on the street, you explore the performance difference between the R1 and Touring packages, pull over, get out of the car, put your hands on the hood, and wait for the officer to arrive."

Motor Trend tested top speed of the fivesome—they also threw in the more expensive Acura NSX—and found the Mazda in fourth place at 163mph, ahead of only the Mitsubishi and behind the Corvette (172.2mph), NSX (171.7mph), and the 300ZX (165mph). But on the 2.5-mile Willow Springs road course, the R1 was the fastest around, turning a 1min 38.3sec fastest lap, followed by the Corvette at 1:39.5, NSX (1:39.9), 3000GT (1:40.3) and 300ZX (1:40.9). On the other hand, given $40,000 to spend, the editors chose the 300ZX as their first choice as their own car, the Corvette and RX-7 "tied for a distant second."

Road & Track used the Streets of Willow Springs for its comparison test, not for lap speed but to test "the best-handling sports cars in America." Racer Danny Sullivan did the driving and describing, saying of the RX-7, "It's like a big Miata, in that it has similar handling characteristics. You have to be careful, though, because if you are hard on the throttle while cornering, it can get light at the rear. But with this turbo setup, the power hit is very consistent. The steering effort is very light. It turned in real well, and once the car is pointed, the tail comes out a bit and the car handles nicely. It doesn't have as much power as some of the others, but I'm really impressed with the engine's kick."

With twelve sports cars to choose from, including the Nissan NX 2000 at one end of the price range and the Porsche 911 Turbo at the other, Sullivan picked the Porsche as his favorite handler, although three of six editors favored the RX-7.

How good was the RX-7 overall? *Motor Trend* didn't announce its "Import Car of the Year" until its February 1993 issue, almost a year after the RX-7's formal introduction, but it gave Mazda's sports car the title anyway. Said the magazine, "The Mazda dramatically advances the level of performance in its class, as evident in both our objective testing and the subjective evaluations by our crew of judges. On every editor's scorecard, the RX-7 ranked in first place."

There were a few complaints. Some thought a few of the interior fittings weren't up to usual Mazda standards, and the snug interior was found unsuitable for the portly. Some didn't like the R1's front and rear spoilers. But if there was a consistent criticism of the RX-7, it was that the ride, particularly that of the R1, was too harsh. Said Douglas Kott in *Road & Track*, "I felt the compromise of this R1 package, in the general region of my kidneys. This is a seriously stiff-riding car, even to gung-ho enthusiasts with a soft spot for hard suspensions." On the other hand, the Touring, with softer shock, tires and suspension bushings, was "an RX-7 suitable for, well, touring. Both versions follow the lay of the road like a cruise missile with terrain-sensing radar, but the Touring car is the kinder, gentler weapon."

The Touring model included a power tilt-and-slide sunroof, leather interior, cruise control, air conditioning, and the Bose "Acoustic Wave" sound system. This last option included a special speaker cabinet that used about half of the RX-7's already meager luggage space.

All U.S.-bound RX-7s came with a "leather-wrapped" steering wheel with a hub mounted airbag. Touring models had switches for the cruise control mounted on the hub for easy access. Cloth seating came on base and R1 models, and all RX-7s had power outside mirrors and windows. AM/FM stereo with

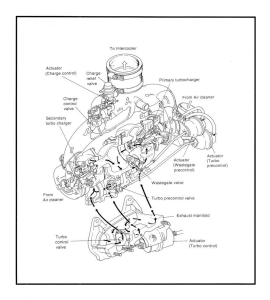

Sequential twin-turbo installation on the 13B-REW engine. The arrows show exhaust flow. Mazda

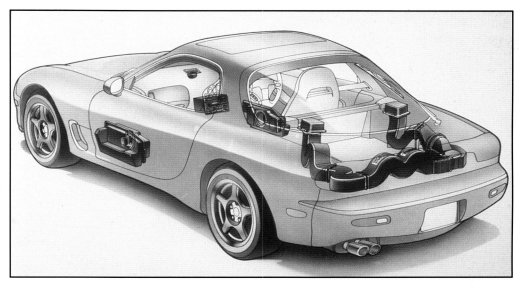

"Bose Acoustic Waveguide" was part of the Touring model's upgrade sound system but robbed much of the already limited trunk space. Bose

cassette was standard with a compact disc player optional. Another stand-alone option was a four-speed automatic transmission. Fuel economy ratings using the EPA test were 17mpg city and 25mpg highway for five speed models and 18mpg city and 24mpg highway for automatic-equipped cars. Note that the coefficient of drag for the base and touring models was 0.29 while the spoilered and bewinged R1 traded low drag for down force, recording a 0.31Cd.

The 1993 RX-7 was, all in all, a critical success. Sales, however, were disappointing. Even though not on sale for a full year—part of 1992's sales were late-1991 models—1992 calendar year sales still totaled only 6,006 RX-7s. It reflected in part a general recession and downturn in Japanese car sales that saw Mazda sales slide even while new versions of the MX-6, 626, and 929 were introduced in the U.S., again each to critical acclaim. While some markets grew by as much as 30 percent (Australia), Mazda's domestic sales were off by sixteen percent and exports were down 6.9 percent. Total volume was down 10.5 percent for fiscal 1993.

Ford Motor Company, which had sold its Flat Rock, Michigan, assembly plant to Mazda Motor Manufacturing (USA) Corp. for the assembly of Mazda 626 and MX-6, as well as the Ford Probe, bought back half of the plant. The organization became a 50-50 joint venture renamed AutoAlliance. The move helped Mazda stay in the black financially.

There would be few changes for the 1994 RX-7. "Refinement" was once again the word used to describe alterations to the model for its second year. Mazda answered criticism of the RX-7's ride by equipping base and Touring models with revalved shock absorbers mounted on softer bushings and substituting urethane rubber bump-stops at the rear and lengthening the front bump-stops. Caster was increased at the front for improved tracking while the power steering pump and gear was revised for a more stable feel. For base and Touring models, the rear anti-roll bar diameter was reduced from 1993's 19.1mm to 15.9mm. As one Mazda representative put it, "they took all the understeer out of it."

The high-performance version was renamed the R-2, the most obvious difference being the change to Z-rated Pirelli P-Zero 225/55R16 tires. Changes similar to those made on other 1994 RX-7s were made on the R-2 as well, although larger diameter shock absorbers were fitted to the R-2 and the larger rear anti-roll bar was retained. For all RX-7s, yellow was removed from the palette of exterior colors, replaced by white.

The 1994 RX-7, however, faced a lingering recession and tough market for any performance or sports car. RX-7 sales for 1993 totaled just 5,062. Despite this, the suggested list price was increased by $2,000 at the beginning of the 1994 model year, even after a $1,100 increase on 1993 models in September. This was in spite of tales of dealers giving huge discounts on RX-7s during 1993. List price for a 1994 base RX-7 was $36,000, surely a bargain for the technology but only for those whose pockets were sufficiently deep. It was a long way in many ways from the 1979 RX-7, 13in wheels, and a $6,395 price tag. Had it really only been fifteen years?

Chapter 15

The Future

Predicting the future is a risky business, at least if outcomes matching predictions mean anything. And putting a prediction in a book which, one hopes, will remain on the shelves for some time certainly puts the prognosticator at more risk than the tabloid fortune-tellers that will be lining the bottom of birdcages next week. One thing is for sure, however, and that is whatever is happening today is a poor predictor of what will happen tomorrow. And projections are virtually worthless: Just because its raining harder at lunchtime than at breakfast doesn't mean it will rain even harder at dinnertime.

That said, it's still interesting to see what future rotaries Mazda is working on so that we can at least speculate on the future, if not predict it. One project Mazda has been working on with a lot of vigor is a hydrogen-fueled rotary. The first public manifestation of that project was a 998cc H-RE10X rotary installed in the HR-X concept vehicle debuted at the Tokyo Motor Show in 1991 and dis-

> As long as there is Mazda there will be the rotary engine.

played at major auto shows around the world. Mazda built two HR-X cars.

The H-RE10X featured separate side air- and hydrogen-intake ports, ceramic apex seals, and the sliding surfaces of the rotor and side-housings were plasma-sprayed with cermet, a metal/ceramic. The engine was rated at 100bhp with a peak torque of 94lb-ft, although acceleration was aided by Mazda's ATCS (automatic torque compensating system), a combination electric motor/generator that boosts torque by as much as 70 percent at 1000rpm. With ATCS, the HR-X could run the quarter-mile in 18.8sec and had a range of 120 miles at a steady 36mph.

The next development was a fleet of rotary hydrogen-powered Miatas. The engines were "standard dimension" naturally aspirated Mazda rotaries, designated H-RE13B, and producing 120bhp and 106lb-ft of torque. An interesting feature of this engine is a camshaft. Hydrogen is admitted via a peripheral port, a poppet valve admitting low-pressure hydrogen. The poppet valve is actuated by what is possibly the only camshaft ever on a rotary engine. The H-RE13B's housing surfaces are hard chrome-plated, and special ceramic apex seals are used.

The Hydrogen Miata, weighing

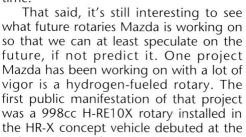

The 1992 Eunos Cosmo, with the first and only multi-rotor production rotary engine. Mazda

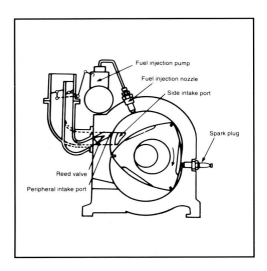

Mazda ROSCO engine of the late-1970s. Mazda

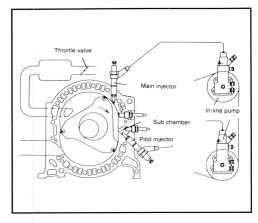

Mazda DISC-II engine of the early-1990s. Mazda

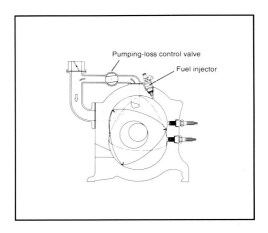

Mazda Miller-cycle rotary, "near-future" design for reduced pumping loss. Mazda

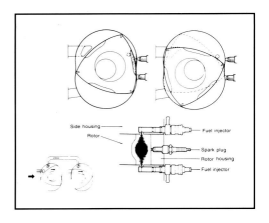

Mazda DISC engine from the late-1980s. Mazda

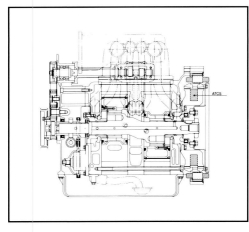

Mazda "green" hydrogen rotary; note the poppet valve and camshaft at the top of the engine. Mazda

2,904lb, can turn the quarter-mile (actually 400m, or 0.24mi) in 17.7sec and has a top speed of 112.8mph. The advantage of hydrogen is that it is a very clean-burning fuel. It theoretically produces only water as its exhaust emission, though in real life partially burned lubricating oil results in carbon monoxide and, from atmospheric nitrogen, oxides of nitrogen, although both are emitted in extremely small quantities.

With only a small catalytic converter, however, the hydrogen rotary can beat California's "Draconian" (Mazda's adjective) ULEV (ultra-low-emission vehicle) rules. The hydrogen rotary is less likely to backfire than a reciprocating hydrogen engine, since in a rotary the volatile hydrogen fuel is not subjected to a hot spark plug during the induction or compression cycles. A disadvantage of hydrogen is its relatively low density of energy content compared to gasoline. Storage is also a problem. The metal hydride fuel tanks of the Hydrogen Miata weigh 770lb; the engine including ancillaries weighs only 457.6lb.

Experimental gasoline rotaries have included ROSCO from the 1970s and DISC from the 1980s. ROSCO was an acronym for ROtating Stratified COmbustion. Unique features of ROSCO were a reed-valve side intake port, an additional side intake port, and direct fuel injection. The direct injection and port layout allowed a rich pocket of charge to be ignited, which burned the leaner mixture following on the trailing side. The side port supplied additional air at higher rpm.

DISC, for Direct-Injection Stratified Charge, evolved in the late-1980s. Two low-pressure injectors were sited on either side of the rotor chamber close to the trailing spark plug. High-speed air injected in the nozzles' sockets aided fuel atomization. Engine speed was controlled by fuel injection, which reduced pumping losses by the absence of a throttle. The engine also required less cooling as combustion took place in a more localized area, which also reduced emissions of unburned fuel (hydrocarbons). But the lower exhaust temperature caused another problem: It wouldn't light up the catalytic converter!

DISC-II solves that problem with an elaborate fuel/air management system. It involves a pilot injector and spark plug in a small subchamber peripheral to the main chamber. The pilot injector sprays through the subchamber to the main chamber, forming a pocket of rich mixture in the main chamber. This pocket is ignited by a flame front from the stable mixture in the subchamber. Meanwhile, the main injector shoots fuel into the main chamber past a conventionally (for a rotary) located spark plug.

And while Mazda put Miller-cycle technology into production in a piston engine in the 1995 Mazda Millenia in the spring of 1994, it was also working with the pumping-loss philosophy with the ro-

Old Name, New Car

The new Cosmo, introduced in 1992, was powered by the 13B-REW or, optionally, the three-rotor 20B-REW. The 20B-REW evolved from the engine used in the one-off three-rotor RX-7 shown to the American press in late-1989 and, similar to that engine, the 20B-REW used a built-up eccentric shaft.

Like the two-rotor 13B-REW, the three-rotor engine is also twin-turbocharged, but is tuned for use in a luxury coupe rather than a sports car. Instead of the RX-7's double Hitachi HT12 turbos, the Cosmo gets a larger HT15 primary turbo. The primary turbo has an "impact" exhaust turbine, the exhaust gases hitting the turbine blades at almost a right angle for maximum initial response. The secondary turbo has a "high-flow" turbine to maximize impeller speed. Additional charge is forced into the combustion chamber by resonance effects in six separate tracts (two for each rotor) and a large plenum.

A novel exhaust feature is a triple-mode muffler system, using two main silencers with four tailpipes. Butterfly valves are opened sequentially at mid and high rpm to minimize noise while producing maximum horsepower. The 20B-REW, as installed in the Cosmo, yields 280bhp JIS at 6500rpm, and 296.8lb-ft of torque at 3,000rpm.

The bad news is that the Cosmo is produced for the home market only, and even before the financial pall that began in 1993, Mazda had no plans to bring the Cosmo to the U.S. Yet the Cosmo, sold by MMC's Eunos sales outlets in Japan, served its purpose for the RX-7 enthusiast, when it served as Mazda's rotary Grand Tourer and allowed the RX-7 to be pure sports car.

Earning the titles of "Mr. Mazda" and "Godfather of the Rotary Engine," Kenichi Yamamoto saw the rotary engine develop from curiosity to high-tech production engine. Yamamoto retired at the end of 1992. Automobile Quarterly Photo and Research Library

tary engine. "Pumping loss" is the energy an engine wastes at closed or small-throttle operation, when trying to draw in air. The Miller-cycle rotary, called "near-future" in Mazda press material, allows a great volume of air to be supplied to the rotor chamber. Any excess air is released via a "late" side port controlled by a "pumping-loss control valve" which recirculates the excess air to the intake port behind the throttle. Direct fuel injection solves the problem of mixture control with a single injector later in the cycle than the "exit port."

Concern that Mazda plans to abandon the rotary should be somewhat allayed by the Miller-cycle rotary, which promises to reduce the engine's thirst for fuel. More importantly, it demonstrates that active research continues with the engine. Mazda senior managing director Michinori Yamanouchi cites the hydrogen rotary and other "future" (his quotes) rotaries "that will carry Mazda through the 1990s and beyond."

Yamanouchi also called Mazda and the rotary engine "inseparably linked." As long as there is Mazda there will be the rotary engine. The RX-7 is virtually defined by the rotary. Are we perhaps at the dawn of another golden age of the rotary engine?

Appendices

Specifications

1979 RX-7 S and GS

PRICES
Base price S	$6,395
Base price GS	$6,995
Options	
Alloy wheels	$250
Air conditioning	$525
3-speed automatic	$355
Sunroof	$275
Calif. emmissions	$75

GENERAL
Curb weight	2,420lb
Wheelbase	95.3in
Track, f/r	55.9/55.1
Length	168.7in
Width	65.9in
Height	49.6in
Ground clearance	6.1in
Fuel cap.	14.5gal, U.S.

ENGINE
Type	twin rotary
Displacement	1146cc/70.0ci
Compression ratio	9.4:1
Output	100bhp (net) @ 6000rpm
Torque	105lb-ft @ 4000rpm
Fuel supply	Two-stage, 4-barrel carburetor

DRIVETRAIN

Transmission	4-speed	5-speed	3-speed
Gear ratios			
1st	3.674	3.674	3.458
2nd	2.217	2.217	1.458
3rd	1.432	1.432	1.000
4th	1.000	1.000	——
5th	——	0.825	——
Final drive ratio	3.909	3.909	3.909

STEERING
Type	recirculating ball
Overall ratio	variable, 17-20:1
Turns, lock to lock	3.7
Turning circle	31.5ft

SUSPENSION
Front, type	independent(1)
Anti-roll bar	23mm
Rear, type	live axle(2)
Anti-roll bar	none (S); 18mm (GS)

BRAKES
Front, type	ventilated disc
diameter	8.94in
thickness	0.71in
Rear, type	Finned drums
diameter	7.87in
thickness	1.26in
Swept area f/r	1037/402sq-in
Tires	165HR-13 (S); 185/70HR-13 (GS)
Wheels	13x5in styled steel (std) 13x5.5in alloy (optional)

EPA Fuel Mileage
5-speed (49-state)	17/28mpg
Auto (49-state)	16/24mpg
5-speed (Calif)	16/27mpg
Auto (Calif)	16/22mpg

1981 RX-7

BASE PRICE
S	$8,595
GS	$9,095
GSL	$10,495

GENERAL

Curb weight	2,345lb (5-speed)
	2,380lb (automatic, 49-state)
Wheelbase	95.3in
Track, f/r	55.9in/55.1in
Length	170.1in
Width	65.7in
Height	49.6in
Ground clearance	5.7in
Fuel cap.	16.6gal, U.S.

ENGINE

Type	twin-rotor, 12A
Displacement	70cc/1146ci
Compression ratio	9.4:1
Output	100bhp (net) @ 6000rpm
Torque	105lb-ft @ 4000rpm
Fuel supply	downdraft two-stage, 4-barrel carburetor

DRIVETRAIN

Transmission	5-speed manual	3-speed automatic
Gear ratios		
1st	3.674	2.458
2nd	2.217	1.458
3rd	1.432	1.000
4th	1.000	——
5th	0.825	——
Reverse	3.542	2.181
Final drive ratio	3.909	3.909

STEERING

Type	recirculating ball
Overall ratio	variable, 17-20:1
Turns, lock to lock	3.7
Turning circle	31.5ft

SUSPENSION

Front, type	independent(1)
Anti-roll bar	23.0mm
Rear, type	live axle(2)
Anti-roll bar	15.0mm

BRAKES

	S and GS	GSL
Front, type	Ventilated disc	same
diameter	8.94in	same
thickness	0.71in	same
Rear, type	finned cast-iron drum	disc
diameter	7.87in	9.29in
thickness	1.26in	0.39in
Swept area, f/r	1,037/402sq-in	1,037/868sq-in

Tires	165HR-13 steel belted radials (S)
	185/70HR-13 steel belted radials (GS and GSL)
Wheels	5.0x13in styled steel (S and GS)
	5.5x13in aluminum alloy (optional, std. on GSL)

EPA Fuel Economy Ratings

49-state	
5-speed manual	21/30mpg
3-speed auto	19/24mpg
California	
5-speed manual	20/30mpg
3-speed auto	19/24mpg

1984 GSL-SE

BASE PRICE	$16,125
POPULAR OPTIONS	
Leather interior	$720
Power steering	$310

GENERAL

Curb weight	2,590lb
Wheelbase	95.3in
Track, f/r	55.9/55.1
Length	170.1in
Width	65.7in
Height	49.6in
Ground clearance	5.7in
Fuel cap.	16.6gal, U.S.

ENGINE

Type	twin rotary
Displacement	1308cc/80ci
Compression ratio	9.4
Output	135bhp (net) @ 6000rpm
Torque	133lb-ft @ 2750rpm
Fuel supply	electronic fuel injection

DRIVETRAIN

Transmission	5-speed manual
Gear ratios	
1st	3.622
2nd	2.186
3rd	1.419
4th	
5th	0.758
Reverse	3.493
Final drive ratio	4.076

STEERING

Type	recirculating ball, power-assist optional
Overall ratio	variable, 17-20:1/15.83:1 (power)
Turns lock to lock	3.4
Turning circle	32.8ft

SUSPENSION

Front, type	independent(1)
Anti-roll bar	23.0mm
Rear, type	live axle(2)
Anti-roll bar	15.0mm

BRAKES

Front, type	ventilated disc
diameter	9.84in
thickness	0.71in
Rear, type	Disc
diameter	10.1in
thickness	0.39in
Swept area, f/r	1,037/868sq-in
Tires	Pirelli 205/60VR14
Wheels	5-1/2x14in aluminum alloy

EPA FUEL ECONOMY RATING 16/23mpg*

*(1985 rating; the EPA revised its mileage calculations for all 1985 models. City ratings are 90 percent of those for 1984, and highway ratings are 78 percent of those for 1984.)

1986 RX-7

PRICES

Two-seater	$11,995
Two-seater GXL	$16,645
2+2	$12,495
2+2 GXL	$17,145

GENERAL

Curb weight	2,625lb (two-seater and 2+2, w/5-speed)
	2,695lb (two-seater w/4-speed automatic)
Wheelbase	95.7in
Track, f/r	57.1/56.7in
Length	168.9in
Width	66.5in
Height	49.8in
Ground clearance	5.9in
Fuel cap.	16.6gal, U.S.

ENGINE

Type	twin rotor
Displacement	1308cc/80ci
Compression ratio	9.4:1
Output	146bhp (net) @ 6500rpm
Torque	138lb-ft @ 3500rpm
Fuel supply	electronic fuel injection

DRIVETRAIN

Transmission	5-speed manual	4-speed automatic
Gear ratios		
1st	3.475	2.841
2nd	2.002	1.541
3rd	1.366	1.000
4th	1.000	0.720
5th	0.711	—
Reverse	3.493	2.400
Final drive ratio	4.100	3.909

STEERING

Type	rack and pinion (power optional)
Overall ratio	20.3 (15.2 power)
Turns, lock to lock	3.5 (2.7 power)
Turning circle	32.2ft

SUSPENSION

Front, type	independent(3)
Anti-roll bar	22mm (24 Sports suspension)
Rear, type	independent(4)
Anti-roll bar	12mm (14 Sports suspension)

BRAKES

	Base	GXL
Power assist std.		
Front, type	ventilated disc	ventilated disc, 4-piston
diameter	9.8in	10.9in
Rear, type	solid disc	ventilated disc
diameter	10.3in	10.7in
Tires	185/70HR14	205/60VR15
Wheels	5.5x14in styled steel/ aluminum alloy opt.	6.0x15in aluminum alloy

EPA FUEL ECONOMY

5-speed manual	17/24mpg
4-speed auto	17/24mpg

1987 RX-7 Turbo

BASE PRICE

1987	$19,345
1988	$21,800 (later $22,750)
10th Anniv.(1988)	$24,650

GENERAL

Curb weight	2,845lb
Wheelbase	95.7in
Track, f/r	57.1/56.7in
Length	168.9in
Width	66.5in
Height	49.8in
Ground clearance	5.9in
Fuel cap.	16.6gal, U.S.

ENGINE

Type	twin rotor turbocharged
Displacement	1308cc/80ci
Compression ratio	8.5:1
Maximum boost	6.2psi
Output	182bhp (net) @ 6500rpm
Torque	183lb-ft @ 3500rpm
Fuel supply	electronic fuel injection

DRIVETRAIN

Transmission	5-speed manual
Gear ratios	
1st	3.483
2nd	2.015

3rd	1.391
4th	1.000
5th	0.762
Reverse	3.288
Final drive ratio	4.100

STEERING
Type	rack and pinion (power optional)
Overall ratio	20.3 (15.2 power)
Turns, lock to lock	3.5 (2.7 power)
Turning circle	32.2ft

SUSPENSION
Front, type	independent(3)
Anti-roll bar	24mm
Rear, type	independent(4)
Anti-roll bar	14mm

BRAKES
Power assist standard, anti-lock optional
Front, type	ventilated disc, 4-piston
diameter	10.9in
Rear, type	ventilated disc
diameter	10.7in
Tires	205/55VR16
Wheels	7.0x16in aluminum alloy

EPA FUEL ECONOMY
5-speed manual	17/23mpg

1988 RX-7 (non-turbocharged models)

BASE PRICES 1988 (mid-year)
SE	$15,480	($16,150)
GTU	$17,350	($18,150)
GXL	$19,160	($20,050)
SE 2+2	$15,980	($16,650)
GXL 2+2	$19,660	($20,550)
Convertible	$20,500	($21,550)

GENERAL
Curb weight	2,625lb (Two-seater w/5-speed)
	2,695lb (Two-seater w/4-speed automatic)
	2,645lb (2+2 w/5-speed)
	2,715lb (2+2 w/4-speed automatic)
	3,003lb (convertible)
Wheelbase	95.7in
Track, f/r	57.1/56.7in
Length	168.9in
Width	66.5in
Height	49.8in
Ground clearance	5.9in
Fuel cap.	16.6gal, U.S.

ENGINE
Type	twin rotor
Displacement	1308cc/80ci
Compression ratio	9.4:1
Output	146bhp (net) @ 6500rpm
Torque	138lb-ft @ 3500rpm
Fuel supply	electronic fuel injection

DRIVETRAIN
Transmission	5-speed	4-speed automatic
Gear ratios		
1st	3.475	2.841
2nd	2.002	1.541
3rd	1.366	1.000
4th	1.000	0.720
5th	0.711	——
Reverse	3.493	2.400
Final drive ratio	4.100*	3.909

*(final drive ratio for convertibles: 3.909:1)

STEERING
Type	rack and pinion, power assist
Overall ratio	15.2
Turns, lock to lock	2.7
Turning circle	32.2ft

SUSPENSION
Front, type	independent(3)
Anti-roll bar	22mm (SE), 24mm (GTU, GXL, Convertible)
Rear, type	independent(4)
Anti-roll bar	12mm (SE, Convertible), 14mm (GTU, GXL)

BRAKES
Power assist standard, ABS optional on GXL

	SE	GTU, GXL, Convertible, ventilated disc, 4-piston
Front, type	ventilated disc	ventilated disc, 4-piston
diameter	9.8in	10.9in
Rear, type	solid disc	ventilated disc
diameter	10.3in	10.7in
Tires	185/70HR14	205/60VR15
Wheels	5-1/2x14in aluminum	6.0x15in aluminum alloy

EPA FUEL ECONOMY
5-speed manual	17/24mpg
4-speed auto	17/23mpg

1989 RX-7

PRICES
Base Prices 1989 (mid-year)
GTU	$17,300
GTUs	$19,600
GXL	$21,600
GXL 2+2	$22,100
Turbo	$22,750
Convertible	$25,950
Options	

Automatic trans.	$750	
Air conditioning	$859	

GENERAL
Curb weight	2,800lb (Two-seater w/5-speed)
	2,844lb (Two-seater w/4-speed automatic)
	2,806lb (2+2 w/5-speed)
	2,851lb (2+2 w/4-speed automatic)
	3,045lb (convertible w/5-speed)
	3,105lb (convertible w/4-speed automatic)
Wheelbase	95.7in
Track, f/r	57.1/56.7in
Length	169.9in
Width	66.5in
Height	49.8in
Ground clearance	5.9in
Fuel cap.	18.5gal, U.S.

ENGINE
Type	twin rotor	twin rotor turbocharged
Displacement	1308cc/80ci	1308cc/80ci
Compression ratio	9.7:1	9.0:1
Output	160bhp(net)@ 7000rpm	200bhp(net)@ 6500rpm
Torque	140lb-ft @ 4000rpm	196lb-ft @ 3500rpm
Fuel supply	Electronic fuel injection	

DRIVETRAIN
Transmission	5-speed	4-speed auto.	5-speed Turbo
Gear ratios			
1st	3.475	2.841	3.483
2nd	2.002	1.541	2.015
3rd	1.366	1.000	1.391
4th	1.000	0.720	1.000
5th	0.697*	—	0.756
Reverse	3.493	2.400	3.288
Final drive ratio	4.100*	3.909*	4.100

*(5th gear for GTU^S is 0.756, final drive ratio is 4.30:1; final drive ratio for convertible automatic is 4.100:1)

STEERING
Type	rack and pinion, power assist
Overall ratio	15.2 (17.2 for GTU)
Turns, lock to lock	2.7 (3.09 for GTU)
Turning circle	32.2ft

SUSPENSION
Front, type	independent(3)
Anti-roll bar	24mm
Rear, type	independent(4)
Anti-roll bar	12mm (Convertible), 14mm (all other)

BRAKES
Power assist standard, ABS optional on GXL

	GTU	GTU^S, GXL, Convertible
Front, type	ventilated disc	ventilated disc, 4-piston
diameter	9.8in	10.9in
Rear, type	Solid disc	Ventilated disc
diameter	10.6in	10.75in
Tires	GTU, GXL, Conv. 205/60VR15	GTU^S, Turbo 205/55VR16
Wheels	GTU, GXL 6.0x15in alloy	Conv. 6-1/2x15in alloy	GTU^S, Turbo 7.0x16in alloy

EPA FUEL ECONOMY
5-speed manual	17/25mpg
4-speed auto	17/24mpg
Turbo	16/24mpg

1991 RX-7

BASE PRICES
Coupe	$19,335
Turbo	$26,555
Convertible	$27,715

GENERAL
Curb weight	2,787lb (Coupe w/5-speed)
	2,831lb (Coupe w/4-speed automatic)
	3,071lb (Convertible w/5-speed)
	3,115lb (Conv. w/4-speed automatic)
	3,003lb (Turbo)
Wheelbase	95.7in
Track, f/r	57.1/56.7in
Length	169.9in
Width	66.5in
Height	49.8in
Ground clearance	5.9in
Fuel cap.	16.6gal, U.S.

ENGINE
Type	Twin Rotor	Twin Rotor Turbo
Displacement	1308cc/80ci	
Compression ratio	9.7:1	9.0:1
Output	160bhp(net)@ 7000rpm	200bhp(net) @ 6500rpm
Torque	140lb-ft @ 4000rpm	196lb-ft @ 3500rpm
Fuel supply	electronic fuel injection	

DRIVETRAIN
Transmission	5-speed	4-speed auto.	5-speed Turbo
Gear ratios			
1st	3.475	2.841	3.483
2nd	2.002	1.541	2.015
3rd	1.366	1.000	1.391
4th	1.000	0.720	1.000
5th	0.697	—	0.719
Reverse	3.493	2.400	3.288
Final drive ratio	4.100*	3.909	4.100

*(final drive ratio for convertible with automatic: 4.100:1)

STEERING		
Type	rack and pinion, power assist	
Overall ratio	17.4:1 (15.2:1 Turbo)	
Turns, lock to lock	3.09 (2.7 Turbo)	
Turning circle	32.2ft	

SUSPENSION
Front, type — independent(3)
  Anti-roll bar — 24mm

Rear, type — independent(4)
  Anti-roll bar — 12mm (Convertible), 14mm (Coupe and Turbo)

BRAKES
Power assist standard, anti-lock std. on Turbo

	Coupe, Conv.	Turbo, option on Coupe
Front, type	ventilated disc	ventilated disc, 4-piston
diameter	10.9in	10.9in
Rear, type	solid disc	ventilated disc
diameter	10.3in	10.75in
Tires	205/60VR15	205/55VR16

Wheels
  Coupe — 6.0x15in alloy
  Convertible — 6-1/2x15in alloy
  Turbo — 7.0x16in alloy

EPA FUEL ECONOMY
5-speed manual — 17/25mpg
4-speed auto — 17/23mpg
Turbo — 16/24mpg

1993 RX-7

BASE PRICES
1993 — $32,900
1994 — $34,000

GENERAL
Curb weight — 2,789lb (base w/5-speed)
2,857lb (base w/automatic)
2,800lb (R1)
2,862lb (Touring w/5-speed)
2,923lb (Touring w/automatic)
Wheelbase — 95.5in
Track, f/r — 57.5/57.5in
Length — 168.5in
Width — 68.9in
Height — 48.4in
Ground clearance — 4.5in
Fuel cap. — 20.0gal. U.S.

ENGINE
Type — twin rotor twin sequential turbocharged
Displacement — 1308cc/80ci
Compression ratio — 9.0:1
Output — 255bhp (bet) @ 6500rpm
Torque — 217lb-ft @ 5000rpm
Fuel supply — electronic fuel injection

DRIVETRAIN

Transmission	5-speed manual	4-speed automatic
Gear ratios		
1st	3.483	3.027
2nd	2.015	1.619
3rd	1.391	1.000
4th	1.000	0.694
5th	0.719	—
Reverse	3.288	2.727
Final drive ratio	4.100	3.909

STEERING
Type — rack and pinion, power assist
Overall ratio — 15.0:1
Turns, lock to lock — 2.9
Turning circle — 35.4ft

SUSPENSION
Front, type — (5)
  Anti-roll bar — 30mm (tubular)

Rear, type — (6)
  Anti-roll bar — 19.1mm (tubular) (17.3mm on 1994 Base and Touring, 15.9mm on 1994 R2)

BRAKES
Power assist, anti-lock standard
Front, type — ventilated disc, four piston
  diameter — 11.6in
Rear, type — ventilated disc
  diameter — 11.6in

Tires — 225/50VR-16 (R1 ZR-rated tires)

Wheels — 8.0x16in squeeze-cast aluminum alloy

EPA FUEL ECONOMY
5-speed manual — 17/25mpg
4-speed auto — 18/24mpg

Suspension Footnotes
(1) Independent MacPherson strut type, with anti-roll bar and tension rods, coil springs
(2) Live axle, 4 trailing arms, Watt linkage, coil springs, anti-roll bar (on some models)
(3) Independent MacPherson strut type, with lower A-arm, anti-roll bar, and coil springs
(4) Independent, Dynamic Tracking Suspension System with trailing arms, Triaxial Floating Hubs, camber control links, coil springs, anti-roll bar
(5) Independent, double-wishbones, coil springs, anti-roll bar
(6) Independent, double-wishbones, coil springs, anti-roll bar

Sales of Mazda Automobiles in the U.S. by Calendar Year

Sales of Mazda Automobiles in the U.S. by Calendar Year*

Year	Total	RX-7
1970	2,305	
1971	19,630	
1972	57,850	
1973	119,003	
1974	61,190	
1975	65,351	
1976	35,383	
1977	50,609	
1978	75,309	19,299
1979	156,533	54,853
1980	161,623	43,731
1981	166,088	43,418
1982	163,638	48,889
1983	173,388	52,514
1984	169,666	55,696
1985	211,093	53,810
1986	222,756	56,243
1987	208,025	38,345
1988	256,050	27,814
1989	263,378	16,249
1990	268,807	9,743
1991	269,699	6,986
1992	280,745	6,006
1993	290,474	5,062

*With few exceptions, model year data is not available, nor has any breakdown by equipment or version (with a few exceptions) been made available by Mazda.

Determining Model Year

Model years can be determined from the serial number/VIN. This number visible on the left front edge of the dash and readable through the windshield. For the 1979 model year it's eleven digits long and begins with SA22C-500025 in 1978 for the 1979 model year. The 1980 model year is eleven digits long as well, beginning with SA22C-558503.

In 1981, the VIN was changed to a seventeen digit number. The tenth digit, an alpha character, indicates the model year, beginning with "B" for 1981. Therefore, the 1982 model year has a "C" in the tenth digit, the 1983, a "D," and so forth. The letters "I," "O" and "Q" are omitted from the series to avoid confusion with the digits 1 and 0. Therefore the 1994 model year has an "R" in the tenth digit.

Finally, the "C" in the eleventh digit indicates the manufacturing plant. The final six digits, beginning with "500000," represent serial production.

Bibliography

Recommended Reading
A number of sources were consulted in writing this book, far more than there was room to list, but the following major sources are recommended for reference, extra reading, and library building.

Ansdale, R.F. *The Wankel RC Engine*. London: Iliffe Books Ltd, 1968. An early treatise on the rotary, very technical.

Dark, Harris Edward. *The Wankel Rotary Engine: Introduction and Guide*. Bloomington, Ind.: Indiana University Press, 1974. A "biography" of the rotary engine before GM quit and the bottom fell out.

Dinkel, John. "Technical Analysis & Driving Impressions: Mazda RX-7." *Road & Track* (May 1978): 43-49, 51-52. Introduction article to new RX-7.

Faith, Nicholas. *Wankel: The Curious Story Behind the Revolutionary Rotary Engine*. New York: Stein and Day, 1975. An extremely detailed recounting of the heyday of the rotary engine.

Fuller, Don. "Mazda 3-Rotor RX-7: Displacement up 50%—with expected results." *Motor Trend* (February 1989): 122-124. Driving impressions of a factory engineering project.

Grable, Ron. "1986 Mazda RX-7: We are talking seriously all new here." *Motor Trend* (November 1985): 57-59, 61, 63, 65, 67-68. Road test.

Hall, Bob. "Mazda RX-7 Pacific Avatar." *AutoWeek* (August 4, 1980): 9. Description and driving impressions of an RX-7 convertible conversion.

Hulett, Burge. "RX-7: Take one for fast relief." *AutoWeek* (January 2, 1984): 10-13, 25. Road test of 1984 RX-7 GSL-SE.

"Humming an Old Favorite." *AutoWeek* (March 1, 1993): 18-21. An "AutoFILE" on the third-generation RX-7.

Ishiwatari, Yasushi. "Mazda RX-7 Turbo GT-X: A quick face-saving fix." *Car and Driver* (December 1984): 34. Driving impressions of the turbocharged Japan-market RX-7.

Lamm, John, and Wally Wyss. "MT Interview: Dr. David Cole." *Motor Trend* (November 1972): 76-77, 109. Talking with Michigan State's rotary doctor when the rotary was hot.

Ludvigsen, Karl. "What Hath Felix Wankel Wrought?" *Motor Trend* (February 1974): 57-60, 65-66. A rotary update in early 1974.

Matras, John. *Illustrated Mazda Buyer's Guide*. Osceola, Wisc.: Motorbooks International, 1994. Focuses on Mazda models from the 1967 Cosmo to the 1993 models.

Mazda Rotary Engine. Hiroshima, Japan. Mazda Motor Corporation, 1993. Press information booklet outlining the history of the Mazda rotary engine, packed with information and with a forward by Kenichi Yamamoto and an introduction by MMC senior managing director Michinori Yamanouchi.

"Mazda RX-7, The First Decade: A test of time: 1979 RX-7, 1988 RX-7 GTU and 1988 RX-7 Turbo." *Road & Track* (August 1988): 68-69, 72-73, 76. A retro-test and test of the GTU and Turbo.

McCluggage, Denise. "Rotary Passion." *AutoWeek* (October 14, 1985): 12-13. Profile of Kenichi Yamamoto.

Norbye, Jan P. *The Complete History of the German Car, 1886 to the Present*. New York: Portland House, 1987. Of course it never mentions the RX-7 nor even Mazda, but it has information and background on German rotary-engined automobiles.

Nunn, Peter. "The New Magic of Mazda." *AutoWeek* (August 21, 1989): 18-20, 24. An interview with Kenichi Yamamoto, late in his career, regarding the future of Mazda and the rotary engine.

Wyss, Wally. "The Engine Detroit Couldn't Ignore." *Motor Trend* (November 1972): 62-64, 66, 68, 70, 72, 74, 76. A history of the Wankel engine written during its heyday.

Yamaguchi, Jack. *RX-7—The New Mazda RX-7 and Mazda Rotary Engine Sports Cars*. New York: St. Martin's Press, 1985. A comprehensive review of the second generation prepared with Mazda Motor Corporation's full cooperation.

Yamaguchi, Jack. *The Mazda RX-7: Mazda's Legendary Sports Car*. Kanagawa, Japan: RING Ltd, 1992. Mazda must have liked the long-time *Road & Track* contributor's effort on the second generation, reprised here for the third-generation RX-7.

The author also consulted the Annual Reports for Toyo Kogyo Co., Ltd., and Mazda Motor Corporation, for the various years in question up through 1993.

Index

Bentele, Max, 10
bibliography, 126
Bonneville Speed Week, 103
Brown, C.R., 26, 27
Citroen, 11, 12
Corvette 2-Rotor, 13
Corvette 4-Rotor, 13
Curtiss-Wright, 10
Fogel, Sidney, 28
Ford Motor Company, 12
Froede, Walter, 10
General Motors, 12, 14, 15, 28
H-RE10X, 117
hydrogen-powered Miata, 117
Katayama, Yoshimi , 71
Kent Racing, 48, 54
Kijima, Takao, 71
Kobayakawa, Takaharu , 86, 107–109
Maebayashi, Jiro, 71

Masuda, Jujiro, 18
Matsuda, Kohei, 28, 32
Matsuda, Tsuneji, 17, 20, 25
Mazda Cosmo, 23
Mazda Familia, 25
Mazda GLC, 28
Mazda Infini IV, 96
Mercedes-Benz C111, 12
Millen, Rod , 43, 54, 55, 57, 66, 67, 82
Nissan, 12
NSU ro80, 11
NSU Spyder, 11
NSU, 9–11, 21
Ohzeki, Hiroshi, 74
Pacific Avatar, 56
Racing Beat, 58, 59, 81, 103
RX-3, 26
RX-4, 26, 27
specifications, 120

St. Yves, Dick, 99
Toyota, 12
Uchiyama, Akio , 31, 32, 70
Wankel, Felix, 9
Yamamoto, Kenichi, 17, 19, 22

Models
Project X605, 31-17
RX-7
    1979–80, 39–43
    1981, 45–49
    1982–83, 51–59
    1984–85, 61–67
    1986, 69-79, 81, 82
    1987, 79, 82, 83
    1988, 83, 85–89
    1989–94, 99–115